Science Laws and Their Applications

Science Laws and Their Applications

By

Kulwant Singh

Cambridge
Scholars
Publishing

Science Laws and Their Applications

By Kulwant Singh

This book first published 2023

Cambridge Scholars Publishing

Lady Stephenson Library, Newcastle upon Tyne, NE6 2PA, UK

British Library Cataloguing in Publication Data
A catalogue record for this book is available from the British Library

ISBN (10): 1-5275-2910-X
ISBN (13): 978-1-5275-2910-6

TABLE OF CONTENTS

PREFACE

Science laws are statements that describe natural phenomena or relationships in the physical world that have been repeatedly observed and confirmed through empirical evidence and experimentation. These laws are based on observations, measurements, and calculations, and are often expressed in mathematical terms.

Examples of well-known scientific laws include:

Newton's Laws of Motion: These laws describe the behavior of objects in motion and explain the relationship between force, mass, and acceleration.

The Law of Conservation of Energy: This law states that energy cannot be created or destroyed, only transformed from one form to another.

Ohm's Law: This law describes the relationship between the voltage, current, and resistance in an electrical circuit.

Boyle's Law: This law explains the relationship between the pressure and volume of a gas at a constant temperature.

The Law of Universal Gravitation: This law describes the relationship between the masses of two objects and the gravitational force between them.

These laws are fundamental to our understanding of the natural world and are the foundation upon which many scientific theories and applications are built. This book describes the various laws used in physical sciences and elaborates on the some of the applications of each of these laws.

The book summarizes the Science Laws at one place, so it will be very useful to the wide range of audience: school students, college students, engineering students, PhD students, scientists and academicians.

1. AMPERE'S LAW

Ampere's law is a fundamental law in electromagnetism, discovered by André-Marie Ampère in the early 19th century. The law relates the magnetic field generated by a steady current to the current density that creates it.

The mathematical expression of Ampere's law in integral form is:

$$\oint \mathbf{B} \times \mathbf{dl} = \mu_0 \mathbf{I}$$

where B is the magnetic field, I is the current passing through the surface enclosed by the closed path dl, and μ_0 is the permeability of free space.

The integral form of Ampere's law is used to calculate the magnetic field around a current-carrying wire or a conductor. The law states that the line integral of the magnetic field around a closed path is proportional to the current enclosed by the path.

The law can also be expressed in differential form, which is useful in the study of electromagnetic waves and in the design of antennas. The differential form of Ampere's law is:

$$\nabla \times \mathbf{B} = \mu_0 \mathbf{J}$$

where J is the current density.

André-Marie Ampère

Ampere's Law

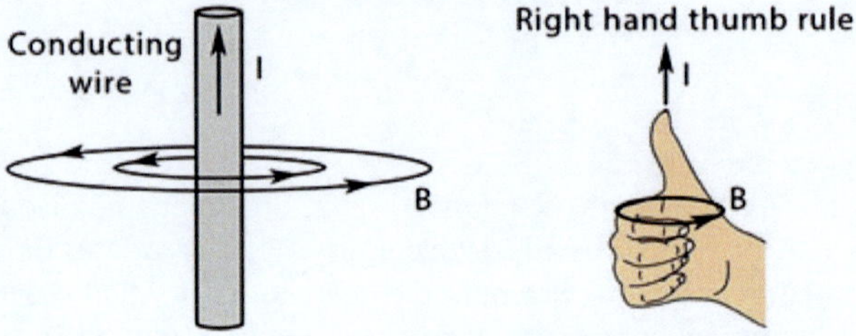

Integral form: $\oint \vec{B}.\vec{dl} = \mu_o I$

Differential form: $\vec{\nabla}\, X\, \vec{B} = \mu_o \vec{J}$

I : Electric curent

B : Magnetic field

μ_o: Permeability of free space

J : Current density

Thumb points in the direction of the electric current and fingers curl around the current indicating the direction of the magnetic field

Applications of Ampere's Law

Ampere's law has many practical applications, including:

- **Design of electromagnets:** Ampere's law is used in the design of electromagnets, which are used in a wide range of applications, including electric motors, MRI (Magnetic Resonance Imaging) machines, and particle accelerators.
- **Magnetic field measurement:** Ampere's law is used in the measurement of magnetic fields generated by currents, such as in the study of magnetic materials and in the design of magnetic sensors.
- **Calculation of magnetic fields:** Ampere's law is used to calculate the magnetic field generated by various current distributions, such as wires, solenoids, and toroids.
- **Study of plasmas:** Ampere's law is used in the study of plasmas, which are important in fusion research and plasma-based devices.

Ampere's law is a crucial principle in electromagnetism, with numerous practical applications in various fields of physics and engineering.

2. ARCHIMEDES' PRINCIPLE

Archimedes' principle is a fundamental principle in physics that relates to buoyancy and the behavior of objects immersed in fluids. It was discovered by the ancient Greek mathematician and scientist Archimedes. Archimedes' principle states that when an object is submerged in a fluid, it experiences an upward buoyant force equal to the weight of the fluid it displaces. In other words, the buoyant force exerted on an object is equal to the weight of the fluid displaced by the object.

This principle can be understood by considering the forces acting on an object submerged in a fluid. The object experiences a downward force due to its weight, which can be thought of as the force of gravity acting on it. At the same time, the fluid exerts an upward force on the object, called the buoyant force. This principle has many practical applications.

Archimedes of Syracuse

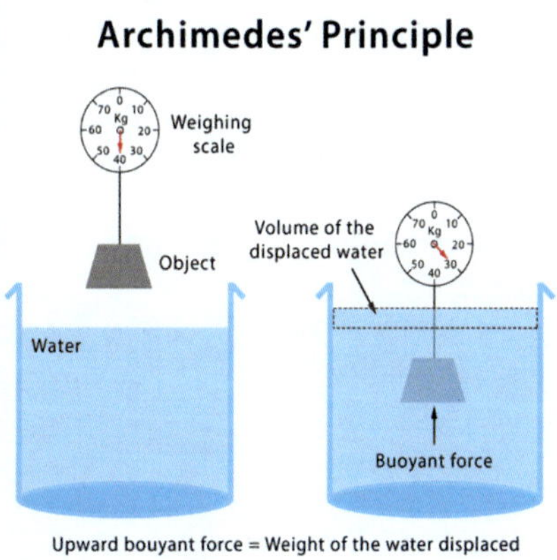

Applications of Archimedes' Principle

Archimedes' principle has many practical applications in a variety of fields. By understanding the principles of buoyancy and fluid dynamics, engineers and scientists can design more effective and efficient systems and devices. Some of the notable applications of Archimedes' principle are:

- **Diving:** Archimedes' principle is important in the sport of diving. Divers use weights and buoyancy compensators to control their buoyancy and remain at a desired depth. Understanding the principles of buoyancy is essential for safe and effective diving.
- **Hot air balloons:** Archimedes' principle is also used in the design and operation of hot air balloons. By heating the air inside the balloon, the air becomes less dense than the surrounding air and creates a buoyant force that lifts the balloon into the air.
- **Density measurements:** Archimedes' principle is used to determine the density of solid objects. By measuring the weight of an object in air and then in water, the volume of water displaced can be calculated, allowing for the calculation of the object's density. This technique will give more accurate results especially for complex shapes and porous materials.
- **Oil drilling:** Archimedes' principle is used in the drilling of oil wells. By injecting a fluid with a density higher than the oil, the oil is displaced upward, allowing it to be pumped to the surface.
- **Design of ships and submarines:** The principle of buoyancy is used extensively in the design and construction of ships and submarines. The shipbuilders use this principle to calculate the buoyant force acting on the ship's hull, and to determine its stability and safety while at sea.
- **Water treatment:** Archimedes' principle is used in water treatment plants to separate solids from liquids. This is done by introducing a coagulating agent into the water, which causes the suspended solids to clump together and settle at the bottom of the tank, leaving the clear water above.
- **Food industry:** Archimedes' principle is used in the food industry to measure the fat content in milk, cheese, and other dairy products. The density of the fat in the milk can be measured using Archimedes' principle, and this information is used to determine the fat content.

3. ARRHENIUS EQUATION

The Arrhenius equation is a mathematical formula that describes the temperature dependence of chemical reactions. It is named after the Swedish chemist Svante Arrhenius, who first proposed it in 1889. The equation relates the rate constant (k) of a chemical reaction to the temperature (T) and the activation energy (E_a) of the reaction:

$$\mathbf{k = A \times exp\ (-E_a / RT)}$$

where A is the pre-exponential factor (sometimes called the frequency factor), R is the gas constant, and exp is the exponential function. The Arrhenius equation shows that the rate constant of a reaction increases exponentially with increasing temperature. It also implies that the activation energy is related to the temperature dependence of the reaction rate, and that a higher activation energy leads to a more temperature-sensitive reaction. The Arrhenius equation is widely used in many branches of science and engineering, including chemistry, physics, materials science, and chemical engineering.

Svante August Arrhenius

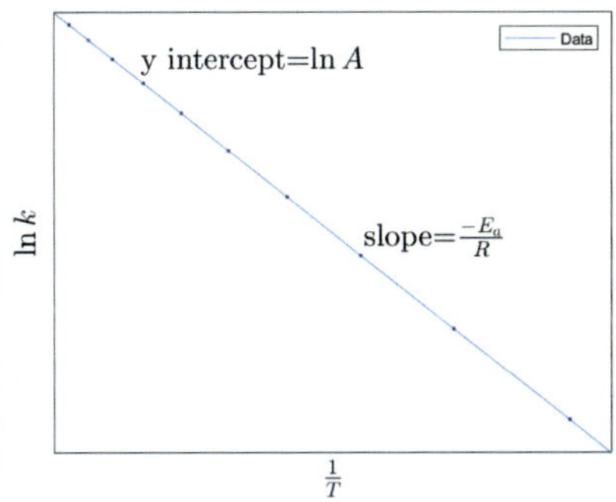

Applications of Arrhenius Equation

The Arrhenius equation is a mathematical equation that relates the rate constant of a chemical reaction to temperature. It is commonly used in the field of chemical kinetics and has several applications in various scientific and engineering fields. Here are some notable applications of the Arrhenius equation:

- **Chemical kinetics:** The Arrhenius equation is primarily used to study the temperature dependence of reaction rates. By measuring reaction rates at different temperatures, the Arrhenius equation can be used to determine the activation energy (Ea) of a reaction and the frequency factor (A) in the rate equation. This information helps in understanding the reaction mechanism and predicting reaction rates at different temperatures.
- **Thermal analysis:** The Arrhenius equation is widely used in thermal analysis techniques such as differential scanning calorimetry (DSC) and thermogravimetric analysis (TGA). These techniques involve measuring changes in sample properties (such as heat flow or weight loss) as a function of temperature. The Arrhenius equation can be applied to analyze the thermal behavior of materials, including decomposition reactions, oxidation, phase transitions, and degradation kinetics.
- **Materials science:** The Arrhenius equation finds applications in materials science and engineering for studying the temperature-dependent properties of materials. It helps determine parameters such as diffusion coefficients, activation energies for diffusion, and thermal stability of materials. These insights are crucial for designing materials with specific properties and understanding their behavior under different thermal conditions. This information is also useful to synthesize various compounds.
- **Pharmaceutical industry:** In pharmaceutical research and development, the Arrhenius equation is utilized to evaluate the stability of drugs and predict their shelf life. By conducting accelerated stability studies at elevated temperatures, the reaction rate constants can be determined using the Arrhenius equation. This information allows for estimating the degradation rate of drugs at typical storage temperatures and establishing appropriate storage conditions.
- **Environmental science:** The Arrhenius equation has applications in environmental science, particularly in modeling and predicting chemical reactions in atmospheric chemistry. By considering the temperature dependence of reaction rates, the Arrhenius equation helps in understanding the kinetics of important processes such as pollutant degradation, atmospheric reactions, and photochemical reactions.
- **Thermal treatment processes:** Industrial processes involving heat treatments, such as metal heat treatment or polymer curing, can benefit from the Arrhenius equation. By understanding the temperature dependence of reaction rates, process parameters can be optimized to achieve desired material properties or efficient processing conditions.

4. AUGER-MEITNER EFFECT

The Auger-Meitner effect, also known as the Auger effect or Auger decay, is a fundamental physical process that occurs in atoms, molecules, and solids. It involves the emission of an Auger electron when an atom undergoes an electronic transition.

The process was first theorized by Lise Meitner in 1922 and later experimentally observed by Pierre Victor Auger in 1923. It is named after both scientists. The Auger-Meitner effect played a crucial role in the development of quantum mechanics and our understanding of atomic and molecular interactions.

The Auger-Meitner effect occurs in atoms or molecules with multiple electron shells. When an inner-shell electron is ionized, typically by an X-ray photon or by electron collision, an electron from an outer shell fills the vacancy. This transition releases energy, which can be transferred to another electron in the system. Instead of emitting a photon, the excess energy is transferred to a third electron, which is subsequently ejected from the atom. This ejected electron is called an Auger electron.

The energy of the emitted Auger electron is characteristic of the atom or molecule involved in the process. By measuring the energy spectrum of Auger electrons, scientists can gain valuable information about the electronic structure and chemical composition of materials.

Lise Meitner **Pierre Victor Auger**

(Credit: French Academy of Sciences – Institut de France)

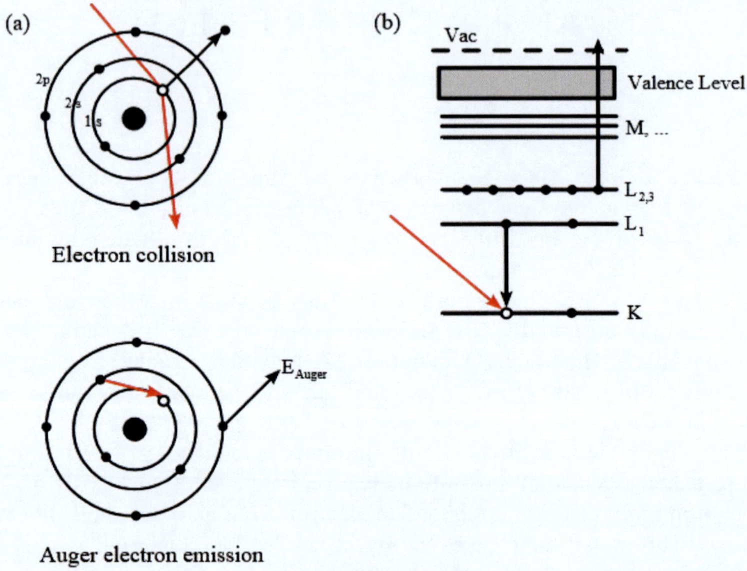

Principle of AES

Two views of the Auger process. (a) illustrates sequentially the steps involved in Auger deexcitation. An incident electron (or photon) creates a core hole in the 1s level. An electron from the 2s level fills in the 1s hole and the transition energy is imparted to a 2p electron which is emitted. The final atomic state thus has two holes, one in the 2s orbital and the other in the 2p orbital. (b) illustrates the same process using X-ray notation, $KL_1L_{2,3}$.

Applications of Auger-Meitner Effect

Auger-Meitner effect or Auger electron spectroscopy (AES), has diverse applications, ranging from surface analysis and semiconductor technology to material science, archaeology, and fundamental research. Its ability to provide detailed information about surface and near-surface properties makes it a valuable tool in numerous scientific and technological disciplines. Here are some notable applications:

- **Surface Analysis:** AES is widely used for surface analysis and characterization of materials. It can provide information about the elemental composition and chemical states of the outermost atomic layers of a material. This is particularly useful in fields such as materials science, nanotechnology, and semiconductor industry, where understanding the surface properties is crucial for optimizing device performance.

- **Thin Film Analysis:** AES is employed for analyzing thin films and multilayer structures. It can determine the thickness and composition of individual layers in a thin film stack, providing insights into the film growth processes and interfaces. This is valuable for applications such as semiconductor device fabrication, optical coatings, and magnetic storage media.
- **Semiconductor device development:** AES is utilized in the research and development of semiconductor devices. It can characterize the dopant distribution, elemental composition, and contamination levels in semiconductor materials and devices. This information helps optimize device performance, reliability, and yield.
- **Surface chemistry studies:** AES is employed to study surface chemical reactions and processes. It can provide insights into adsorption, desorption, and surface reactions occurring at the atomic and molecular level. This is important for understanding catalysis, corrosion, surface passivation, and other surface-related phenomena.
- **Material quality control:** AES can be used for quality control and assurance in various industries. It enables rapid and non-destructive analysis of material surfaces, ensuring that the desired composition and purity levels are maintained. This is applicable in industries such as metallurgy, electronics, pharmaceuticals, and automotive.
- **Archaeology and art conservation:** AES finds applications in the analysis of cultural heritage materials. It can identify the elemental composition of artifacts, paintings, and historical objects without causing significant damage. This aids in the authentication, restoration, and preservation of cultural artifacts.
- **Fundamental research:** The Auger-Meitner effect is also studied for its fundamental significance in understanding electron dynamics and interactions in atoms, molecules, and solids. It provides insights into electron energy levels, relaxation processes, and electronic structure. This knowledge contributes to advancements in quantum mechanics and the theoretical understanding of matter.

5. AVOGADRO'S HYPOTHESIS

Avogadro's hypothesis states that "equal volumes of gases, at the same temperature and pressure, contain the same number of molecules." In other words, if you have two containers of gas that are at the same temperature and pressure, and they have the same volume, they will contain the same number of molecules. This hypothesis is also known as Avogadro's law.

Avogadro's law is a gas law that states that at a constant temperature and pressure, the volume of a gas is directly proportional to the number of molecules or atoms present in the gas. This law is named after Amedeo Avogadro, an Italian scientist who first proposed it in 1811.

Mathematically, Avogadro's law can be expressed as

$$V = k \times n$$

where V is the volume of the gas, n is the number of gas molecules or atoms, and k is a proportionality constant. This law is only valid for ideal gases, which are gases that behave according to certain idealized assumptions, such as having no intermolecular forces and occupying no volume themselves.

Avogadro's law is important in many areas of science and engineering, including chemistry, physics, and materials science. It is often used in calculations involving the behavior of gases, such as determining the molar volume of a gas or the number of moles of a gas in a given volume.

Amedeo Avogadro

Applications of Avogadro's Hypothesis

The applications of Avogadro's hypothesis are numerous and include:

- **Stoichiometry:** Avogadro's hypothesis is used to calculate the stoichiometry of chemical reactions, which involves the ratios of reactants and products. By knowing the number of molecules in a given volume, it is possible to calculate the amount of substance (moles) present in that gas.

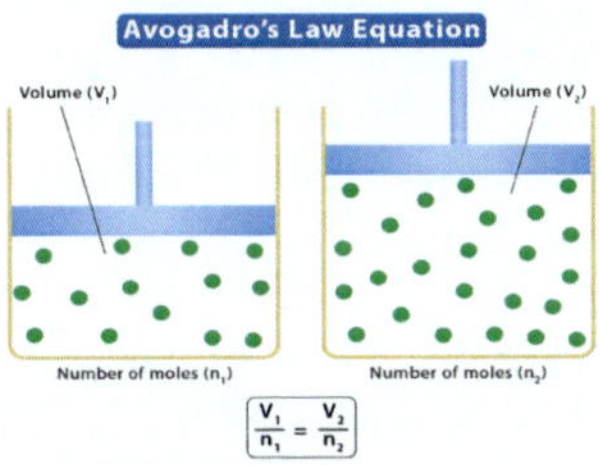

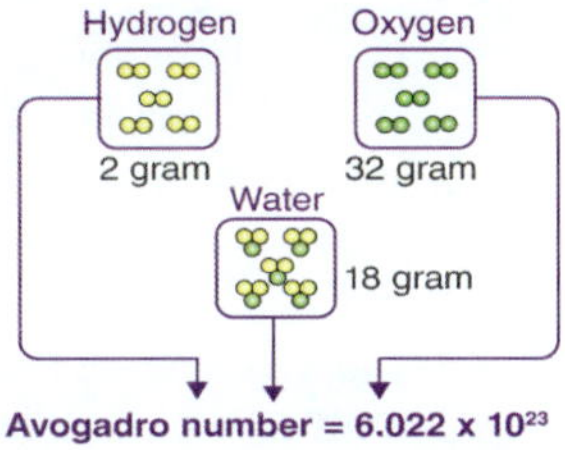

- **Ideal gas law:** Avogadro's hypothesis describes the behavior of gases under certain conditions. The ideal gas law relates the pressure, volume, temperature, and number of molecules of a gas.
- **Atomic and molecular mass:** Avogadro's hypothesis is used to determine the atomic and molecular masses of elements and compounds. By measuring the mass of a given volume of gas, and knowing the number of molecules, it is possible to calculate the mass of each individual molecule.
- **Gas density:** Avogadro's hypothesis is used to calculate the density of gases. By knowing the mass of a given volume of gas, and the number of molecules in that volume, it is possible to calculate the density.
- **Gas mixtures:** Avogadro's hypothesis is used to calculate the composition of gas mixtures. By knowing the number of molecules of each gas in a given volume, it is possible to calculate the proportion of each gas in the mixture.
- **Reaction rates:** Avogadro's hypothesis is used to calculate the rate of chemical reactions that involve gases. By knowing the number of molecules in a given volume of gas, and the rate at which those molecules react, it is possible to calculate the overall rate of the chemical reaction.
- **Molecular collisions:** Avogadro's hypothesis is used to study the behavior of molecules in gases. By knowing the number of molecules in a given volume of gas, and their average speed, it is possible to calculate the rate of molecular collisions, which is an important factor in many chemical and physical processes.

6. BEER-LAMBERT LAW

The Beer-Lambert Law, also known as the Beer-Lambert-Bouguer Law, is a relationship that describes the absorption of light by a material. The law states that the amount of light absorbed by a material is directly proportional to the concentration of the material and the distance that the light travels through the material. The mathematical expression of the Beer-Lambert Law is:

$$A = \varepsilon\, l\, c$$

where A is the absorbance of the material, ε is the molar absorptivity or extinction coefficient of the material, l is the distance that the light travels through the material (known as the path length), and c is the concentration of the material.

August Beer

Johann Heinrich Lambert

Pierre Bouguer

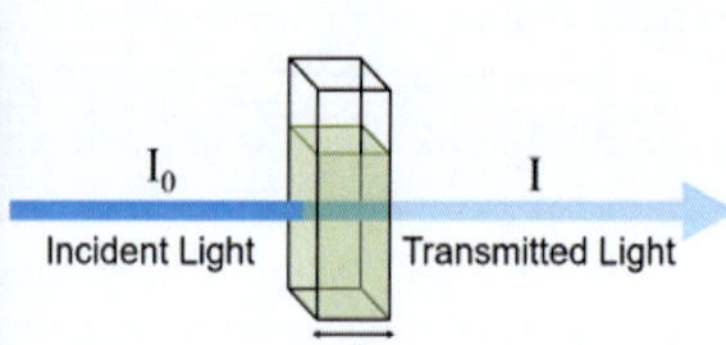

Applications of Beer-Lambert Law

Beer-Lambert Law is a fundamental relationship in the study of light-matter interactions, with many practical applications in science, engineering, and medicine. Here are some of the most common applications:

- **Spectroscopy:** The Beer-Lambert Law is used in spectroscopy to measure the concentration of a substance in a solution. Spectroscopy is the study of the interaction between light and matter, and is used in a wide range of fields, including chemistry, physics, and biology.
- **Chemical analysis:** The Beer-Lambert Law is used in chemical analysis to measure the concentration of a substance in a sample. This can be useful in determining the purity of a substance or in monitoring chemical reactions.
- **Environmental monitoring:** The Beer-Lambert Law is used in environmental monitoring to measure the concentration of pollutants in air or water. This can help to identify sources of pollution and to monitor the effectiveness of pollution control measures.
- **Medical diagnostics:** The Beer-Lambert Law is used in medical diagnostics to measure the concentration of substances in biological samples, such as blood or urine. This can be useful in diagnosing diseases and monitoring the effectiveness of treatments.
- **Quality control:** The Beer-Lambert Law is used in quality control to measure the concentration of different components in a product, such as in the pharmaceutical industry where it is used to ensure the correct concentration of active ingredients in medicines.
- **Forensics:** The Beer-Lambert Law is used in forensic science to identify and quantify different compounds in samples, such as blood or urine, to help solve crimes or accidents.
- **Agriculture:** The Beer-Lambert Law is used in agriculture to measure the concentration of different nutrients, such as nitrogen and phosphorus, in soil samples to determine the health of the soil and the suitability for different crops.
- **Astronomy:** The Beer-Lambert Law is used in astronomy to study the absorption of light by interstellar dust and gas, allowing scientists to study the composition and properties of the interstellar medium.

7. BELL'S THEOREM

Bell's theorem is a fundamental principle in quantum mechanics that demonstrates the non-locality of entangled particles and the limits of classical physics. It was proposed by physicist John Stewart Bell in 1964 and has since been experimentally verified.

The theorem states that any theory based on local realism (i.e., the idea that physical properties exist independently of observation and that information cannot be transmitted faster than the speed of light) must violate certain inequalities, known as Bell inequalities, when applied to measurements of entangled particles.

Entangled particles are pairs of particles that are created in such a way that their quantum states are correlated, even if they are separated by large distances. When one particle is measured, the state of the other particle is instantaneously determined, regardless of the distance between them.

Bell's theorem shows that this instantaneous correlation between the particles cannot be explained by local realism, as it would require information to be transmitted faster than the speed of light. Instead, it suggests that quantum mechanics is a non-local theory, meaning that the properties of particles can be correlated across space without any direct interaction.

The experimental verification of Bell's theorem has significant implications for our understanding of the nature of reality and the limits of classical physics. It also has practical applications in quantum computing and quantum communication, as entangled particles can be used to transmit information in a secure and efficient way.

John Stewart Bell
(Credit: https://cds.cern.ch/record/1823937)

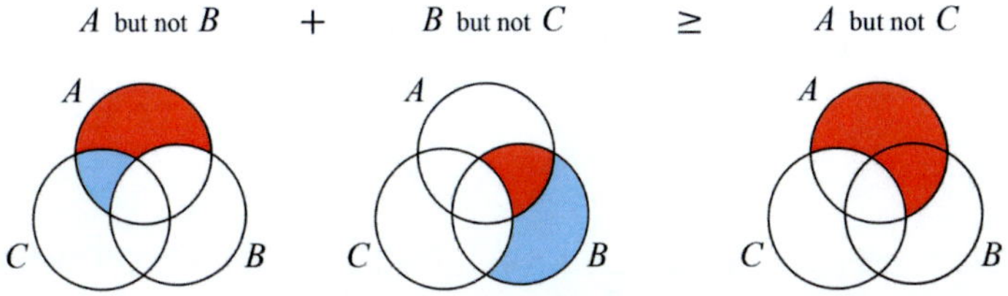

Applications of Bell's Theorem

Bell's theorem has had a significant impact on our understanding of quantum mechanics and has opened up new possibilities for technology and science. Its applications in quantum cryptography, quantum teleportation, and quantum computing have the potential to revolutionize information technology and bring about new advances in fields such as medicine, finance, and materials science. Some of the most common applications include:

- **Quantum cryptography:** Bell's theorem provides the theoretical foundation for quantum cryptography, a method of secure communication that relies on the principles of quantum entanglement. By using entangled particles, quantum cryptography can ensure that information is transmitted in a completely secure and un-hackable way.
- **Quantum teleportation:** Bell's theorem is also essential for quantum teleportation, a process that allows the transmission of quantum information from one location to another. By using entangled particles, quantum teleportation can transfer the quantum state of one particle to another particle, even if they are separated by large distances.
- **Quantum computing:** Bell's theorem has significant implications for quantum computing, a rapidly developing field that aims to build computers that can solve certain problems much faster than classical computers. The non-locality of entangled particles plays a crucial role in quantum computing algorithms, and Bell's theorem helps to explain how these algorithms work.
- **Foundations of quantum mechanics:** Bell's theorem has led to a deeper understanding of the foundations of quantum mechanics, including the role of entanglement and the limits of classical physics. It has also inspired new theoretical frameworks for understanding quantum mechanics, such as Bohmian mechanics and many-worlds interpretation.

8. Bernoulli's Principle

Bernoulli's principle, named after Swiss mathematician Daniel Bernoulli, states that as the speed of a fluid increases, its pressure decreases. This principle is derived from the conservation of energy for a fluid flowing through a pipe or channel, and it applies to both liquids and gases. The principle states that in a streamline flow of an inviscid (non-viscous) and incompressible fluid, the total mechanical energy per unit volume (sum of pressure energy, kinetic energy, and potential energy) remains constant along a streamline. The equation for Bernoulli's principle is expressed as:

$$P + 1/2\ \rho\ v^2 + \rho\ g\ h = \text{constant}$$

where,

P is the pressure exerted by the fluid,

ρ (rho) is the density of the fluid,

v is the velocity of the fluid,

g is the acceleration due to gravity,

h is the height (elevation) of the fluid above a reference point.

It's important to note that Bernoulli's principle assumes certain ideal conditions, such as steady flow, incompressibility, and negligible viscous effects. In practical situations, where these assumptions may not hold, additional factors such as fluid viscosity and losses due to friction must be considered.

Daniel Bernoulli

(Credit: https://www.hmb.ch/en/museums/objects-in-the-collection/details/s/portraet-des-daniel-bernoulli/)

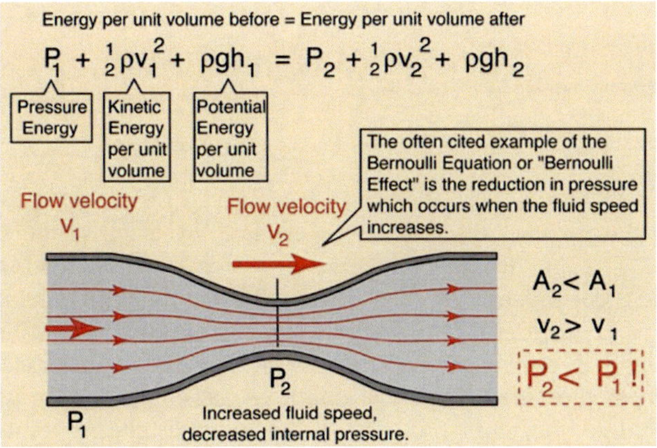

Applications of Bernoulli's Principle

Bernoulli's principle is a fundamental principle in fluid dynamics that describes the relationship between fluid speed and pressure. Its applications are widespread, from aviation and fluid mechanics to automotive design and sports equipment. The practical applications of Bernoulli's principle include:

- **Aviation:** Bernoulli's principle is essential to the understanding of aerodynamics, which is the study of how air moves around objects. The principle explains why airplane wings generate lift, by creating a region of low pressure above the wing as air flows over it faster than beneath it.
- **Fluid mechanics:** The principle is also useful in fluid mechanics, which is the study of the behavior of fluids in motion. It can help explain why water flows faster through a narrower pipe than a wider one.
- **Automotive industry:** Bernoulli's principle is used in the design of carburetors and fuel injectors in combustion engines. These components use the principle to create a pressure difference to atomize fuel for combustion.
- **Ventilation:** Bernoulli's principle is also used in the design of ventilation systems, such as air ducts and HVAC systems, to regulate air flow and pressure.
- **Sports equipment:** Bernoulli's principle is applied in the design of sports equipment such as golf balls and tennis balls, which have dimples and fuzz respectively, to create turbulence in the air flow and reduce drag.

9. Biot-Savart Law

The Biot-Savart law is a fundamental law in electromagnetism that describes the magnetic field generated by a steady electric current. It was discovered independently by Jean-Baptiste Biot and Félix Savart in 1820.

The law states that the magnetic field at any point in space, due to a current-carrying wire or a conductor, is directly proportional to the magnitude of the current and the length of the conductor, and inversely proportional to the square of the distance between the point and the conductor. The direction of the magnetic field is given by the right-hand rule, which states that if you point your right thumb in the direction of the current, then the curl of your fingers gives the direction of the magnetic field.

The mathematical expression of the Biot-Savart law for the magnetic field dB at a point P due to a small segment of wire carrying current I is:

$$dB = (\mu_0 / 4\pi)(I \times dL \times \sin\theta) / r^2$$

where μ_0 is the permeability of free space, dL is the infinitesimal length of the wire segment, r is the distance from the wire to point P, and × denotes the cross product.

The Biot-Savart law is widely used in various fields of physics and engineering, such as electromagnetics, magneto-hydrodynamics, and plasma physics, to calculate the magnetic field due to various current distributions.

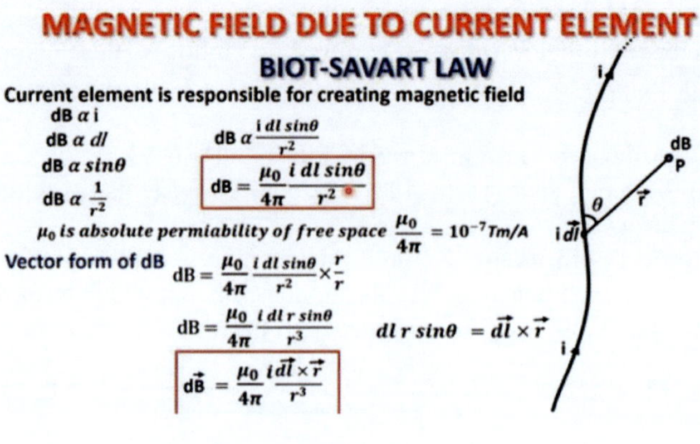

Jean-Baptiste Biot **Félix Savart**
(Credit: https://alchetron.com/F%C3%A9lix-Savart)

Applications of Biot-Savart Law

Biot-Savart law is a crucial tool in the study and application of electromagnetism and plays a significant role in various areas of physics and engineering, some of which are:

- **Calculation of magnetic field:** The Biot-Savart law is used to calculate the magnetic field at any point in space due to a current-carrying wire or a conductor. This is particularly useful in the design and analysis of various electromagnetic devices such as motors, generators, transformers, and magnetic sensors.
- **Current-carrying loops:** The law can be applied to calculate the magnetic field produced by a current-carrying loop, which is useful in the design of electromagnets and in the study of magnetic materials.
- **Magnetic field due to a solenoid:** The Biot-Savart law can be used to calculate the magnetic field produced by a solenoid, which is a device that consists of a coil of wire used to generate a magnetic field.
- **Study of plasmas:** The law is used in plasma physics to calculate the magnetic field generated by a plasma, which is important in understanding the behavior of plasmas in fusion reactors and other plasma-based devices.
- **Magnetic resonance imaging (MRI):** The Biot-Savart law is used in the design and operation of MRI machines, which use magnetic fields to generate images of internal body structures.
- **Magnetic levitation:** The law is used in the design of magnetic levitation trains and other transportation systems that use magnetic fields to levitate and propel vehicles.

10. BOGOLIUBOV HIERARCHY

The Bogoliubov hierarchy, also called BBGKY hierarchy (Bogoliubov–Born–Green–Kirkwood–Yvon hierarchy) is a mathematical framework / set of equations describing the dynamics of a system of a large number of interacting particles. The BBGKY hierarchy, is used to describe the time evolution of correlation functions in a classical or quantum many-particle system. The hierarchy provides a systematic way to study the dynamics and statistical properties of the system by considering correlations between different numbers of particles. This formal theoretic result is named after Nikolay Bogolyubov, Max Born, Herbert S. Green, John Gamble Kirkwood, and Jacques Yvon.

In the classical case, the BBGKY hierarchy consists of a set of coupled equations that describe the time evolution of the one-particle distribution function, the two-particle distribution function, the three-particle distribution function, and so on. Each equation involves correlations between a specific number of particles, and the equations are coupled together through these correlations. The hierarchy provides a way to track how the correlations between particles evolve over time.

The equation for an s-particle distribution function (probability density function) in the BBGKY hierarchy includes the (s+1)-particle distribution function, thus forming a coupled chain of equations. s-particle distribution functions were introduced in classical statistical mechanics by J. Yvon (Yvon 1935, 235). The equations were first introduced by the physicist Nikolay Bogoliubov in 1947 as a way to describe the ground state properties of a Bose-Einstein condensate (BEC), a state of matter that occurs at extremely low temperatures where a large number of bosonic particles occupy the same quantum state.

In the quantum case, the BBGKY hierarchy applies to the reduced density matrices instead of distribution functions. The hierarchy consists of a similar set of equations, but now expressed in terms of the reduced density matrices of different particle numbers. The equations describe the time evolution of the one-particle reduced density matrix, the two-particle reduced density matrix, the three-particle reduced density matrix, and so on.

The BBGKY hierarchy is a powerful tool for studying many-particle systems, particularly in statistical mechanics and quantum many-body theory. However, it becomes increasingly challenging to work with higher-order equations in practice because of the increasing complexity of correlations and computational demands. Therefore, approximations and truncations are often employed to make the hierarchy tractable, such as the truncation to the two-particle level known as the BBGKY equation or the introduction of closure approximations.

Overall, the BBGKY hierarchy provides a systematic framework to study the dynamics and statistical properties of classical and quantum many-particle systems by considering correlations between different numbers of particles. It allows for the investigation of various phenomena, such as equilibration, transport properties, and phase transitions in complex systems.

Nikolay Bogoliubov

(Credit: https://commons.wikimedia.org/w/index.php?curid=44167690)

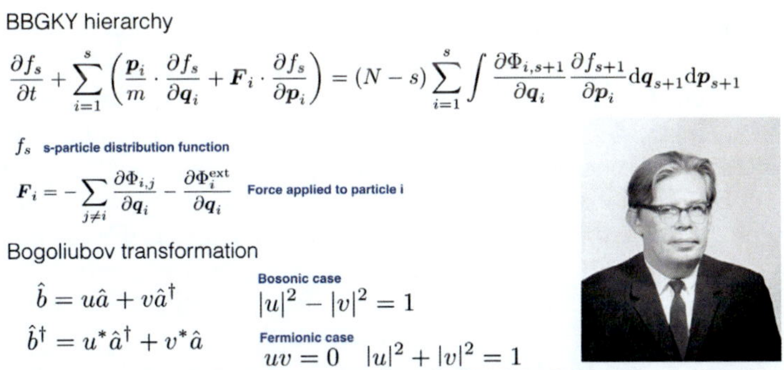

(Credit: https://pbs.twimg.com/media/E9S7uC-XsAEq-b-.jpg:large)

Applications of Bogoliubov Hierarchy

Bogoliubov hierarchy is a mathematical tool used to study the properties of interacting many-body quantum systems. It provides a systematic way of approximating the dynamics of a quantum system by breaking it down into simpler sub-systems. The hierarchy consists of a set of equations that describe the time evolution of the density matrix of the system.

- **Condensed matter physics:** Bogoliubov hierarchy is widely used in the study of Bose-Einstein condensates, which are systems of ultra-cold atoms that exhibit macroscopic quantum behavior. The hierarchy is used to study the collective excitations of the condensate, which are known as Bogoliubov modes. These modes play a crucial role in determining the properties of the condensate, such as its stability, coherence, and response to external perturbations.
- **High energy physics:** In high energy physics, Bogoliubov hierarchy is used to study the dynamics of the quark-gluon plasma, which is a state of matter that is believed to have existed in the early universe, shortly after the Big Bang. The hierarchy is used to study the collective excitations of the plasma, which are known as quasiparticles. These quasiparticles play a crucial role in determining the properties of the plasma.
- **Quantum optics:** Bogoliubov hierarchy is also used in the study of quantum optical systems, such as cavity QED and quantum dots. The hierarchy is used to study the interaction between the light and matter, and to calculate the spectral properties of the system.
- **Cold atoms:** Bogoliubov hierarchy is used to study the properties of cold atom systems, such as the Bose-Hubbard model and the Fermi-Hubbard model. The hierarchy is used to study the dynamics of the system, and to calculate the properties of the ground state and the excited states.
- **Quantum information:** In quantum information theory, Bogoliubov hierarchy is used to study the dynamics of quantum information systems, such as quantum channels and quantum circuits. The hierarchy is used to study the decoherence of the quantum system, and to calculate the properties of the quantum state after it has undergone a quantum operation.

Bogoliubov hierarchy is a powerful tool for studying the collective behavior of interacting bosons in a variety of physical systems. Its applications are not limited to BECs and can be extended to other systems of interacting bosons, such as polariton condensates and exciton-polariton condensates.

11. BOGOLIUBOV TRANSFORMATION

In theoretical physics, the Bogoliubov transformation, also known as the Bogoliubov–Valatin transformation, was independently developed in 1958 by Nikolay Bogolyubov and John George Valatin for finding solutions of Bose-Einstein condensate (BEC) theory in a homogeneous system. It is a mathematical technique used in the study of interacting many-body quantum systems, particularly in the context of superfluidity and superconductivity.

The Bogoliubov transformation is a change of basis for the quantum field operators that describe the particles in the system. The transformation is designed to diagonalize the Hamiltonian of the system in terms of quasiparticle excitations, which are collective excitations of the system that can be thought of as a combination of particles and holes.

The transformation involves a linear combination of the creation and annihilation operators for the particles in the system, where the coefficients of the combination depend on the properties of the system, such as the energy spectrum and the coherence length.

The resulting transformed operators describe the quasiparticle excitations of the system, and the Hamiltonian can be expressed in terms of these operators. This allows for a simpler and more intuitive description of the system, and can be used to calculate the properties of the system, such as the critical temperature and the behavior of the order parameter.

It is a key tool in the study of superfluidity and superconductivity, and has also been applied in the study of other quantum phenomena, such as Bose-Einstein condensation and quantum Hall effects.

The Bogoliubov transformation involves expressing the creation and annihilation operators, known as Bogoliubov operators. The general form of the Bogoliubov transformation for bosonic operators can be written as:

$$\mathbf{a_k = u_k\, \alpha_k + v_k\, \alpha_{-k}^\dagger}$$

$$\mathbf{a_k^\dagger = u_k^*\, \alpha_k^\dagger + v_k^*\, \alpha_{-k}}$$

where a_k and $a_k^\dagger$ are the original annihilation and creation operators, respectively, and α_k and $\alpha_k^\dagger$ are the Bogoliubov annihilation and creation operators. The coefficients u_k and v_k are complex numbers that determine the transformation. The Bogoliubov transformation satisfies the commutation relations for the creation and annihilation operators. It also preserves the bosonic commutation relations for the new operators:

$$\mathbf{[\alpha_k, \alpha_{k'}^\dagger] = \delta_{kk'}}$$

$$[\alpha_k, \alpha_{k'}] = [\alpha_k^\dagger, \alpha_{k'}^\dagger] = 0$$

The coefficients u_k and v_k can be determined by solving the equations of motion or by imposing additional conditions, such as the requirement that the transformed Hamiltonian is diagonal in terms of the Bogoliubov operators.

Nikolay Bogoliubov
(Credit: https://en.wikipedia.org/w/index.php?curid=33768082)

Applications of Bogoliubov Transformation

Bogoliubov transformation is a powerful mathematical tool widely used in various fields of physics, including condensed matter physics, nuclear physics, and quantum field theory. Here are some of its applications:

♦ **Superconductivity:** The Bogoliubov transformation is used to describe the behavior of electrons in a superconductor. By transforming the Hamiltonian into a diagonal form, the transformation reveals the existence of pairs of electrons (Cooper pairs) that are responsible for the superconductivity.

♦ **Bose-Einstein condensation:** The Bogoliubov transformation is used to study the behavior of Bose-Einstein condensates, a state of matter where a large number of bosons occupy the same quantum state.

♦ **Nuclear physics:** The Bogoliubov transformation is used in nuclear physics to study the properties of atomic nuclei. The transformation helps to describe the collective motion of nucleons in the nucleus.

♦ **Quantum field theory:** In quantum field theory, the Bogoliubov transformation is used to relate different sets of creation and annihilation operators, which describe different particles in a system. This transformation helps to describe the interactions between different particles and their excitations.

12. BOLTZMANN DISTRIBUTION

The Boltzmann distribution is a probability distribution that describes the relative populations of different energy states in a system at thermal equilibrium. It is based on the assumption that the particles in the system are in random thermal motion and are subject to the laws of statistical mechanics.

The Boltzmann distribution is given by the equation:

$$P_{Sb}/P_{Sa} = (N_b/N_a) = g_b/g_a * e^{(-(E_b-E_a)/kT)}$$

where P_{Sb} and P_{Sa} are the probabilities of occupying states b and a, respectively, N_b and N_a are the number of particles occupying states b and a, respectively, g_b and g_a are the degeneracy (the number of available states at a given energy level) of states b and a, respectively, E_b and E_a are the energy levels of states b and a, respectively, T is the temperature of the system, and k is the Boltzmann constant.

This equation shows that the probability of occupying a higher energy state (b) relative to a lower energy state (a) decreases exponentially with increasing energy difference (ΔE) between the two states and increasing temperature (T) of the system. The Boltzmann constant (k) relates the energy of a system to its temperature, and is approximately equal to 1.38 × 10^-23 joules per kelvin.

The Boltzmann distribution is a fundamental concept in statistical mechanics and is used to describe the behavior of many physical systems, including gases, liquids, and solids. It provides a basis for understanding phenomena such as phase transitions, chemical reactions, and diffusion.

Ludwig Eduard Boltzmann

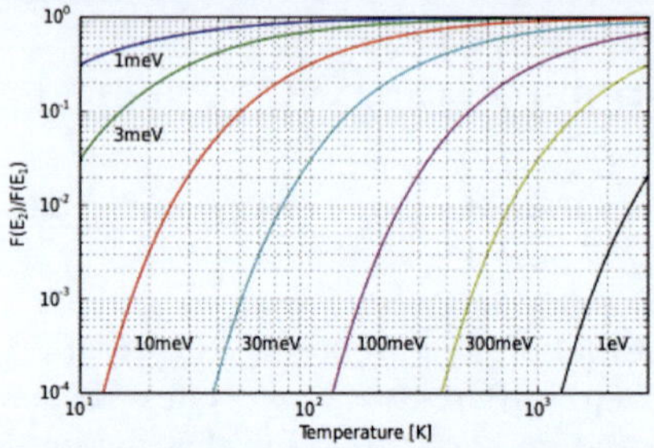

Boltzmann distribution (Sonium. 2015. "Boltzmann distribution graph". Accessed March 30, 2015)

Applications of Boltzmann Distribution

Boltzmann distribution is a fundamental concept that has a wide range of applications in various fields of science and engineering. The Boltzmann distribution, also known as the Gibbs distribution, is a probability distribution that describes the distribution of particles in a system at thermal equilibrium. Some of the most common applications are:

- **Statistical mechanics:** The Boltzmann distribution is the cornerstone of statistical mechanics, that explains the behavior of macroscopic systems based on the properties of their constituent particles. In statistical mechanics, the Boltzmann distribution is used to calculate the probability of a particle being in a particular energy state.

- **Thermodynamics:** The Boltzmann distribution is also used in thermodynamics to calculate the average energy of a system in thermal equilibrium. It can be used to determine the temperature of a system by measuring the average energy of its constituent particles.

- **Molecular biology:** The Boltzmann distribution is used in molecular biology to calculate the probability of a protein or RNA molecule adopting a particular conformation. This is important because the function of these molecules is often dependent on their shape.

- **Astronomy:** The Boltzmann distribution is used in astronomy to calculate the distribution of velocities of particles in a star or galaxy. This information can be used to determine the mass of the object and its distance from Earth.

- **Plasma physics:** The Boltzmann distribution is used in plasma physics to calculate the distribution of charged particles in a plasma. This is important because plasmas are used in many technological applications.

13. BOLTZMANN EQUATION

The Boltzmann transport equation is a fundamental equation in statistical mechanics that describes the behavior of a gas at the molecular level. It was first proposed by Austrian physicist Ludwig Boltzmann in the late 19th century. It provides a microscopic description of how particles, such as electrons or phonons, transport and distribute themselves in a system under the influence of external forces and collisions.

The equation itself is an integro-differential equation that describes the time evolution of the distribution function f(r,p,t), which gives the probability of finding a particle with position r, momentum p, and time t. The Boltzmann transport equation takes into account the effects of collisions between particles, which can cause changes in momentum and redistribute the particles in phase space.

The Boltzmann transport equation can be written as:

$$\partial \mathbf{f}/\partial \mathbf{t} + \mathbf{v}\cdot\partial \mathbf{f}/\partial \mathbf{r} + \mathbf{F}\cdot\partial \mathbf{f}/\partial \mathbf{p} = (\partial \mathbf{f}/\partial \mathbf{t})_{\mathbf{coll}}$$

where v is the velocity vector, F is the external force acting on the particles, and $(\partial f/\partial t)_{coll}$ represents the collision term accounting for interactions and scattering events.

Solving the Boltzmann transport equation allows us to study various transport properties in a system, such as electrical conductivity, thermal conductivity, and diffusion coefficients. It provides insights into the behavior of particles and their distribution in response to external perturbations, providing a connection between microscopic dynamics and macroscopic transport phenomena.

The Boltzmann equation is a highly nonlinear partial differential equation and is difficult to solve exactly. However, it is an essential tool for understanding the behavior of gases, especially at high pressures and temperatures where classical mechanics breaks down and quantum mechanics must be used.

Applications of Boltzmann Equation

The Boltzmann transport equation finds wide application in fields such as condensed matter physics, semiconductor physics, plasma physics, and kinetic theory. Different approximation techniques and methods, such as the relaxation time approximation or the Boltzmann equation with the Grad method, are often used to simplify or solve the Boltzmann transport equation in specific cases.

♦ **Modeling of rarefied gases:** The Boltzmann equation provides a theoretical basis for understanding the behavior of rarefied gases, such as those found in outer space or in high vacuum environments. It is also used in the design of spacecraft and satellites, as well as in the development of vacuum technologies.

♦ **Understanding of transport phenomena:** The Boltzmann equation is used to study transport phenomena in gases, such as thermal conduction and viscosity. These phenomena are important in a wide range of engineering applications, from heat transfer in engines to the design of micro-electromechanical systems (MEMS).

♦ **Development of kinetic theory:** The Boltzmann equation is the foundation of kinetic theory, which is a branch of statistical mechanics that studies the behavior of large numbers of interacting particles. Kinetic theory has applications in many fields, including astrophysics, plasma physics, and materials science.

♦ **Development of Monte Carlo methods:** Monte Carlo methods are a class of numerical techniques used to simulate complex physical systems. They are widely used in fields such as computer science, finance, and engineering. The Boltzmann equation has been used as the basis for the development of Monte Carlo methods for simulating the behavior of gases.

♦ **Semiconductor physics:** The Boltzmann equation is used to model the behavior of electrons and holes in semiconductors. This is important for the design of electronic devices, such as transistors and solar cells.

♦ **Plasma physics:** The Boltzmann equation is used to study the behavior of plasmas, which are ionized gases that are important in many applications, including fusion reactors and plasma processing.

♦ **Astrophysics:** The Boltzmann equation is used to study the behavior of gases in stars, galaxies, and other astronomical objects. This is important for understanding the formation and evolution of the universe.

Boltzmann equation is a powerful tool for understanding the behavior of gases and has numerous applications in many fields of science and engineering. Its ability to model the behavior of individual molecules and particles has made it an essential tool for researchers studying a wide range of physical and chemical phenomena.

14. Born's Law

Born's law or Born's rule, named after the physicist Max Born, is a fundamental principle in quantum mechanics that describes the probability of finding a particle at a particular position in space.

Born's law states that the probability density of finding a particle or a quantum system in a specific state is proportional to the square of the magnitude of the wave function of the particle at that location. In other words, if $\Psi(x)$ is the wave function of a particle, the probability density of finding the particle at position x is given by $|\Psi(x)|^2$.

This means that the wave function of a particle contains all the information about the probability of finding the particle at any given location. The wave function is a complex-valued function that describes the quantum state of the particle, and can be used to calculate a wide range of physical quantities, including the energy spectrum, momentum distribution, and spin of the particle. A wave function ψ for a single structure less particle in space position (x, y, z) implies that the probability density function p for a measurement of the particle's position at time t_0 is:

$$p(x, y, z, t_0) = |\psi (x, y, z, t_0)|^2$$

Born's law is a key principle in quantum mechanics and is used in many applications of the theory, including in the study of atoms, molecules, and solid-state materials. It also plays a crucial role in the interpretation of the results of quantum mechanical experiments, where the wave function of a particle is used to calculate the probability of the particle being detected at a particular position or with a particular energy or momentum.

Max Born

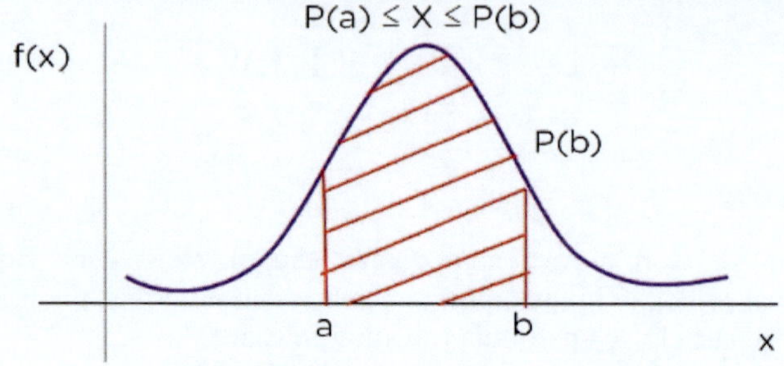

Probability Density Function

Applications of Born's Law

Born's law, also known as the Born rule, is a fundamental concept in quantum mechanics that is widely used in various applications, including:

- **Scattering:** Born's law is used to describe the scattering of particles off a potential barrier or a target. It provides a way to calculate the probability of a particle being scattered in a particular direction, given the properties of the target and the incident particle.
- **Atomic physics:** Born's law is used to calculate the probability of an electron being found in a particular location around an atom's nucleus. This information is used to construct electron density maps of atoms and molecules, which are used in various fields, including chemistry, materials science, and drug design.
- **Solid-state physics:** Born's law is used to calculate the potential energy of a crystal lattice, which provides insight into the properties of materials, including their electrical conductivity, thermal conductivity, and mechanical strength.
- **Quantum optics:** Born's law is used to calculate the probability of a photon being scattered by an object, such as a crystal or a molecule. This information is used to design optical devices, such as filters, lenses, and detectors.
- **Quantum computing:** Born's law is used to calculate the probability of a qubit being in a particular state after undergoing a quantum operation. This information is used in the design of quantum algorithms and quantum error-correction protocols.

15. BOYLE'S LAW

Boyle's law is a fundamental principle in physics that describes the relationship between the pressure and volume of a gas at a constant temperature. It states that the pressure of a gas is inversely proportional to its volume, when the temperature and the amount of gas remain constant.

Mathematically, Boyle's law can be expressed as:

$$P_1 V_1 = P_2 V_2$$

where P_1 and V_1 are the initial pressure and volume of a gas, and P_2 and V_2 are the final pressure and volume, respectively.

Robert Boyle

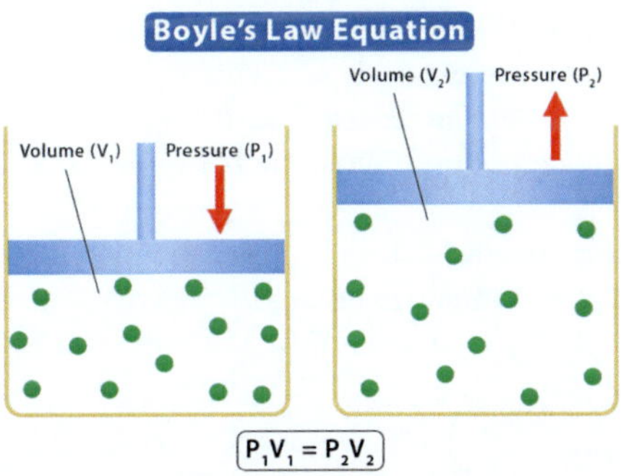

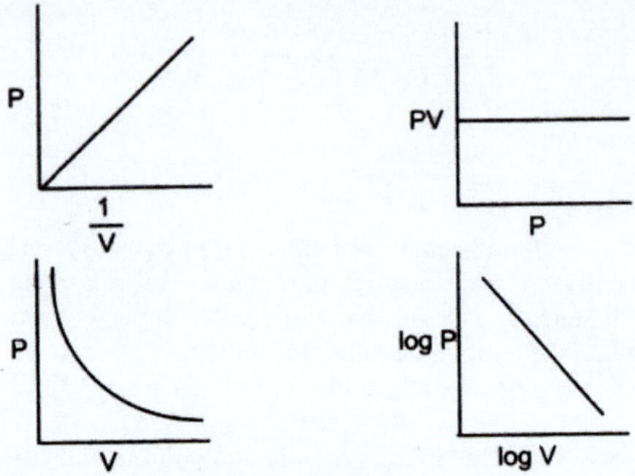

Applications of Boyle's Law

Boyle's law is a fundamental principle in physics that has many applications in various fields of science and engineering. It provides a powerful tool for understanding the behavior of gases under different conditions, and has enabled the development of many important technologies, such as scuba diving equipment and chemical reactors. Boyle's law has many applications in physics and engineering, including:

- **Gas laws:** Boyle's law is one of the gas laws, along with Charles's law and Gay-Lussac's law. These laws describe the behavior of gases under various conditions, and are used in the design of many engineering systems.
- **Breathing:** Boyle's law is used to describe the mechanics of breathing. When a person inhales, the volume of their lungs increases and the pressure inside their lungs decreases. This allows air to flow into the lungs.
- **Scuba diving:** Boyle's law is used in scuba diving to describe the behavior of gases in the human body. As a diver descends into deeper water, the pressure on their body increases, and the volume of air in their lungs decreases. This can cause problems, such as the bends, if not managed properly.
- **Chemistry:** Boyle's law is used in chemistry to describe the behavior of gases in chemical reactions. By controlling the pressure and volume of gases, chemists can control the rate and extent of chemical reactions.

16. BRAGG'S LAW

Bragg's law is a principle of X-ray crystallography that describes the relationship between the wavelength of X-rays, the angle of incidence, and the distance between the planes of atoms in a crystal lattice. It is named after William Lawrence Bragg and his father William Henry Bragg, who developed the law in 1912. Although Laue is credited with the discovery of diffraction.

Bragg's law states that the X-rays will be diffracted by a crystal lattice at a specific angle when the path difference between the X-rays scattered from adjacent crystal planes is an integer multiple of the wavelength of the X-rays. Mathematically, the law is expressed as:

$$2\,d\,\sin\theta = n\,\lambda$$

where d is the distance between adjacent crystal planes, θ is the angle of incidence of the X-rays, λ is the wavelength of the X-rays, and n is an integer.

William Henry Bragg

William Lawrence Bragg

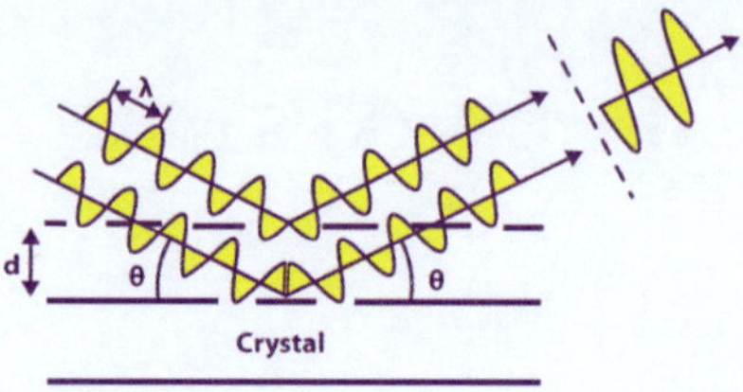

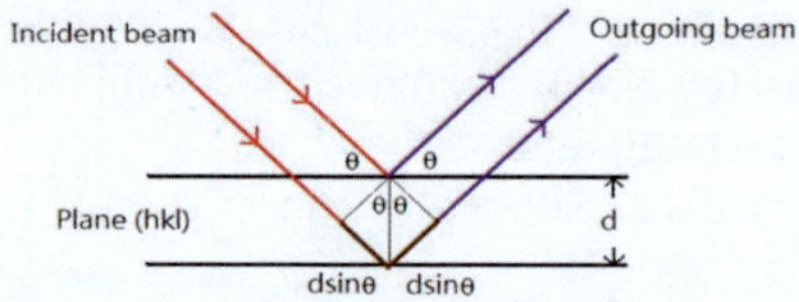

When the total path difference $2d\sin\theta$ equals to $n\lambda$, diffraction occurs.

Bragg's Law: $2d\sin\theta = n\lambda$

n: order of reflection

Applications of Bragg's Law

Bragg's law has important applications in X-ray crystallography and other areas of science and engineering, including:

- **Structure determination of crystals:** Bragg's law is used to determine the structure of crystals by analyzing the diffraction pattern produced by X-rays passing through the crystal. By measuring the angle and intensity of the diffracted X-rays, scientists can determine the distance between the crystal planes and the arrangement of atoms within the crystal.
- **Analysis of materials:** Bragg's law is used to analyze the structure of materials, including metals, semiconductors, and polymers. By analyzing the diffraction pattern produced by X-rays passing through a material, scientists can determine the crystal structure of the material.
- **Non-destructive testing:** Bragg's law is used in non-destructive testing techniques, such as X-ray diffraction and neutron diffraction, to analyze the structure of materials without damaging them.
- **Study of proteins and enzymes:** Bragg's law is used in the study of proteins and enzymes to determine their crystal structures. By analyzing the diffraction pattern produced by X-rays passing through a protein crystal, scientists can determine the three-dimensional structure of the protein, which is important for understanding its function and for designing drugs to target it.

17. BREWSTER'S LAW

Brewster's Law is a principle of optics that describes the relationship between the polarization of light and the angle of incidence on a surface. It states that when a beam of light is incident on a surface at a specific angle, known as the Brewster angle, the reflected light is completely polarized, with the electric field vector of the reflected light perpendicular to the plane of incidence.

The Brewster angle is given by the equation:

$$\tan \theta_B = n_2 / n_1$$

where θ_B is the Brewster angle, n_1 is the refractive index of the medium in which the incident light is traveling, and n_2 is the refractive index of the medium on the other side of the surface.

David Brewster

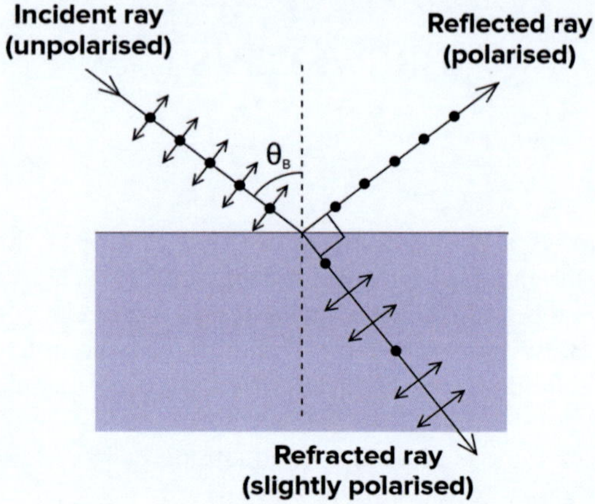

Applications of Brewster's Law

Brewster's Law has several important applications in optics and engineering, including:

- **Polarization of light:** Brewster's Law is used to produce polarized light. By reflecting a beam of light at the Brewster angle, engineers can produce light that is completely polarized, with the electric field vector perpendicular to the plane of incidence.
- **Glare reduction:** Brewster's Law is used to reduce glare from reflective surfaces, such as windows and water. By applying an anti-reflection coating to the surface at the Brewster angle, engineers can reduce the amount of reflected light and minimize glare.
- **Optical coatings:** Brewster's Law is used in the design of optical coatings, such as anti-reflection coatings and polarizing filters. By understanding the relationship between the angle of incidence and the polarization of light, engineers can design coatings that have specific optical properties.
- **Study of optics:** Brewster's Law is fundamental to the study of optics and is used in many areas of research, including the study of thin films, surface plasmon resonance, and the behavior of light in complex media. By understanding the principles of Brewster's Law, scientists can better understand the behavior of light and its interaction with matter.

18. BRUUN RULE

The Bruun Rule is a widely used empirical relationship in coastal engineering that provides an estimate of the maximum run-up height of waves on a beach or coastal structure during a storm event.

The rule was developed by Per Moller Bruun in the 1950s, and is based on the observation that the maximum run-up height of waves is proportional to the square root of the wave height and the water depth. Mathematically, the Bruun Rule is expressed as:

$$R_{max} = \alpha \sqrt{(H\,d)}$$

where R_{max} is the maximum run-up height, H is the wave height, d is the water depth, and α is a dimensionless empirical constant that depends on the characteristics of the beach or coastal structure. The value of α typically ranges from 0.05 to 0.08 for sandy beaches and up to 0.15 for steep rocky shores.

The Bruun Rule is a simple and useful tool for estimating the potential impact of storm waves on coastal structures, such as seawalls, jetties, and breakwaters. It can also be used to design coastal structures to withstand the maximum run-up height of waves during extreme storm events. However, it should be noted that the Bruun Rule is an empirical relationship and may not always accurately predict the maximum run-up height of waves, especially for complex coastal geometries and wave conditions.

Per Moller Bruun

(Credit: https://hydraulicians.en-academic.com/100/BRUUN_Per)

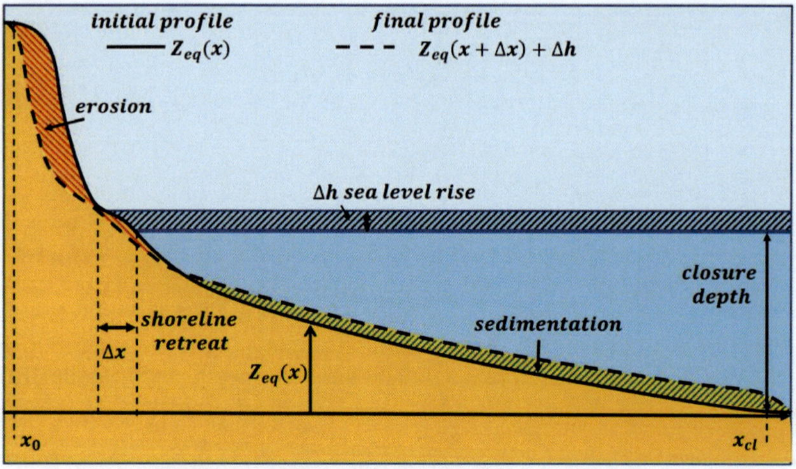

The Bruun Rule: Shoreface shift in response to sea-level rise

(Dronkers, J. 2022. Bruun rule for shoreface adaptation to sea-level rise. Accessed December 11, 2022)

Applications of Bruun Rule

♦ Bruun Rule is a simple empirical formula used to estimate shoreline erosion or accretion caused by the propagation of waves in the littoral zone (the zone where the ocean meets the coast). The rule states that the amount of shoreline recession or accretion (in meters) is roughly equal to 20 times the height of the waves (in meters) multiplied by the square root of the slope of the beach. It assumes that the beach slope is constant, the sediment is uniform and the wave energy is constant.

♦ This formula is useful for preliminary assessments of shoreline change and can help in designing coastal protection measures. However, it should be noted that the Bruun Rule has limitations and should be used with caution as it does not account for the variability of the wave climate, sediment characteristics and other factors that can affect shoreline change.

19. Buys Ballot's Law

Buys Ballot's law, named after the Dutch meteorologist Christophorus Buys Ballot, is an empirical law in atmospheric science that describes the relationship between wind direction and atmospheric pressure.

The law states that if a person stands with their back to the wind in the northern hemisphere, the area of low pressure will be on their left, and the area of high pressure will be on their right. Conversely, if a person stands with their back to the wind in the southern hemisphere, the area of low pressure will be on their right, and the area of high pressure will be on their left.

This law is a consequence of the Coriolis effect, which causes the wind to deflect to the right in the northern hemisphere and to the left in the southern hemisphere due to the Earth's rotation. As a result, air flows around low pressure systems in a counter clockwise direction in the northern hemisphere, and in a clockwise direction in the southern hemisphere.

Buys Ballot's law is a useful tool for predicting the movement of weather systems and is often used in weather forecasting. By analyzing the wind direction and pressure patterns, meteorologists can determine the location and movement of low pressure systems, which are typically associated with stormy weather, and high pressure systems, which are typically associated with fair weather.

Christophorus Buys Ballot

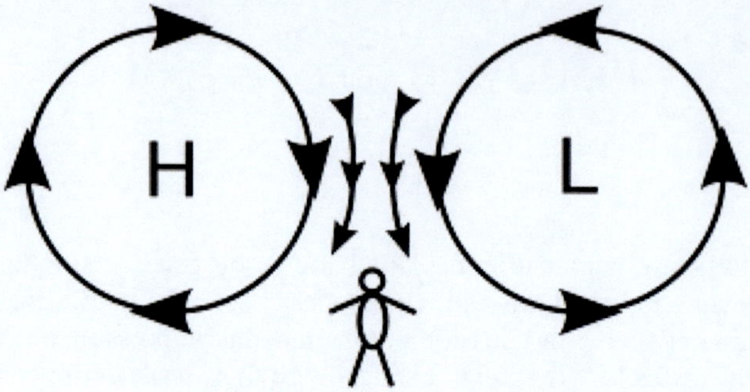

Applications of Buys Ballot's Law

♦ Buys Ballot's law is a meteorological law that states that if an observer stands with his back to the wind in the Northern Hemisphere, low pressure is to the observer's left, and high pressure is to his right. Similarly, in the Southern Hemisphere, low pressure is to the observer's right, and high pressure is to his left.

♦ The law is particularly useful in weather forecasting, as it allows forecasters to make quick predictions about the movement of weather systems. By observing the direction of the wind and using Buys Ballot's law, forecasters can predict the location of high and low-pressure systems and the likely weather patterns associated with them.

♦ In addition to its use in weather forecasting, Buys Ballot's law has also been used in aerodynamics to help engineers understand the movement of air around a wing or other aerodynamic surface. By understanding the direction of the air pressure gradients around a surface, engineers can design more efficient and effective aerodynamic shapes.

20. BYERLEE'S LAW

Byerlee's law, named after the geophysicist Don L. Anderson Byerlee, is an empirical relationship that describes the frictional strength of rocks as a function of the applied pressure. The law states that the frictional strength of rocks increases with increasing pressure, according to a linear relationship. Mathcmatically, Byerlee's law can be expressed as:

$$\tau = \mu \, p + \sigma \, n$$

where τ is the shear stress (frictional strength) of the rock, μ is the coefficient of friction, p is the applied pressure, σ is the effective normal stress (the force acting perpendicular to the plane of shear), and n is a constant that depends on the properties of the rock.

Byerlee's law is important in the study of earthquakes and fault mechanics, as it helps to explain the behavior of rocks in the Earth's crust under tectonic stresses. The law suggests that rocks near a fault may undergo a transition from static friction (where the shear stress required to cause movement is higher than the applied stress) to dynamic friction (where the shear stress required to cause movement is lower than the applied stress) as the applied pressure increases.

The law also has practical applications in rock mechanics and engineering, where it is used to estimate the strength of rocks and to design structures, such as tunnels and dams, that will be subjected to high pressures. However, it should be noted that Byerlee's law is an empirical relationship and may not accurately predict the behavior of all types of rocks under all conditions.

Don L. Anderson Byerlee

(Credit: Bill Youngblood for the California Institute of Technology. https://commons.wikimedia.org/wiki/File:Don_L._Anderson_portrait_photo.jpg)

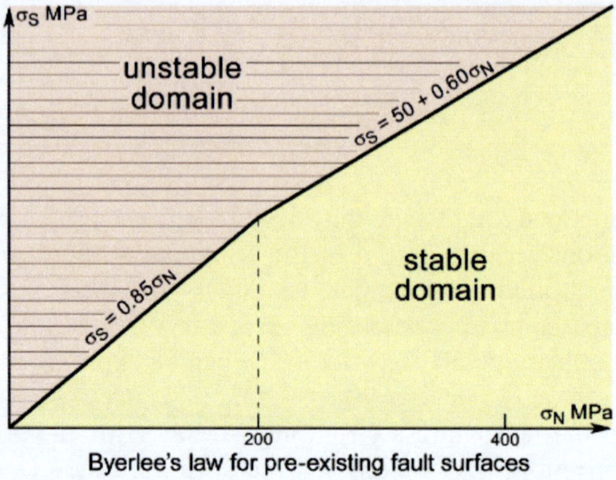

Byerlee's law for pre-existing fault surfaces (Burg 2018, 60)

Applications of Byerlee's Law

Byerlee's law is an empirical relationship that describes the frictional strength of rocks under a range of pressures and temperatures. This relationship is widely used in geophysics and engineering geology to estimate the strength of rocks at depth in the Earth's crust, and to model the behavior of faults and earthquakes.

Some specific applications of Byerlee's law include:

- **Earthquake prediction:** Byerlee's law provides a way to estimate the shear strength of rocks in the Earth's crust, which is a key parameter in earthquake prediction models.
- **Reservoir engineering:** In petroleum engineering, Byerlee's law is used to predict the behavior of rock formations in oil and gas reservoirs under different conditions of pressure and temperature.
- **Geotechnical engineering:** Byerlee's law is used to design underground excavations and tunnels, as well as to analyze the stability of slopes and embankments.
- **Mining engineering:** Byerlee's law is used to predict the strength of rock masses in underground mines, and to design support systems to prevent rock falls and collapses.

Byerlee's law is an important tool for understanding the mechanical properties of rocks and predicting their behavior under a wide range of conditions.

21. CARNOT'S THEOREM

Carnot's theorem is a fundamental principle of thermodynamics that describes the maximum efficiency that a heat engine can achieve. Carnot's theorem is a principle in thermodynamics that relates to the maximum efficiency of a heat engine, which is a device that converts thermal energy into mechanical energy. It was developed by French physicist Sadi Carnot in 1824.

Carnot's theorem states that the maximum efficiency of a heat engine is dependent on the temperature difference between the hot and cold reservoirs it operates between, and that this efficiency is independent of the working substance used in the engine. In other words, the efficiency of a heat engine can be improved by increasing the temperature difference between the hot and cold reservoirs, but it can never exceed a certain limit, which is determined solely by the temperature of the reservoirs.

Carnot's theorem can be expressed mathematically using the Carnot cycle, which is a theoretical cycle that describes the operation of an ideal heat engine. The efficiency of the Carnot cycle is given by the equation:

$$\textbf{Efficiency} = (T_1 - T_2) / T_1$$

where T_1 is the temperature of the hot reservoir, and T_2 is the temperature of the cold reservoir. This equation shows that the efficiency of the engine can be increased by increasing the temperature difference between the hot and cold reservoirs, but it can never exceed $1 - T_2/T_1$, which is the maximum possible efficiency of a heat engine operating between those two temperatures.

Sadi Carnot

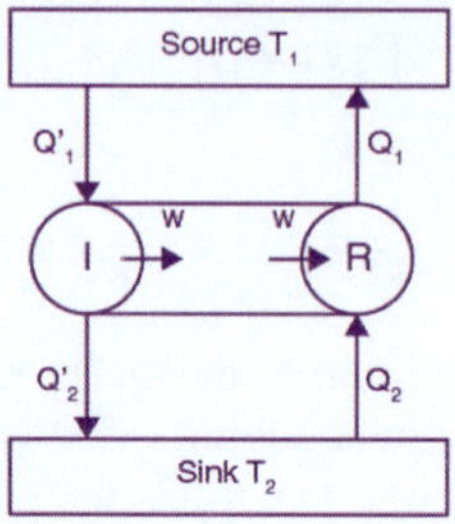

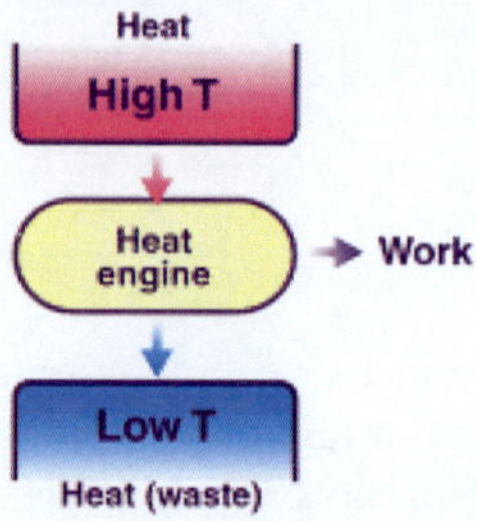

Applications of Carnot's Theorem

It has a number of important applications in various fields of science and engineering. Here are some of the applications of Carnot's theorem:

- **Design of heat engines:** Carnot's theorem is used in the design of heat engines, which convert thermal energy into mechanical work. It provides a theoretical upper limit on the efficiency of a heat engine, and is used to optimize the design of engines to maximize their efficiency.
- **Refrigeration and air conditioning:** Carnot's theorem is used in the design of refrigeration and air conditioning systems. It is used to calculate the theoretical maximum efficiency of the system, and to optimize the design of the system to achieve this maximum efficiency.
- **Solar energy and electricity production:** Carnot's theorem is used in the design of solar energy systems, which convert solar radiation into electricity or thermal energy. It is used to optimize the design of solar energy systems to maximize their efficiency. It is very useful to estimate the efficiency of the conversion of heat energy into electricity in power plants.
- **Study of phase transitions:** Carnot's theorem is used in the study of phase transitions, such as the transition between liquid and vapor phases. It provides a theoretical framework for understanding the thermodynamics of phase transitions and for predicting their behavior.
- **Study of quantum thermodynamics:** Carnot's theorem has been extended to the study of quantum thermodynamics, which is the study of the thermodynamics of small quantum systems. It is used to understand the thermodynamics of quantum systems and to develop new technologies, such as quantum computers.

Carnot's theorem provides a fundamental principle for understanding the maximum efficiency of heat engines, and is used to optimize the design of systems that convert thermal energy into other forms of energy.

22. CASIMIR EFFECT

The Casimir Effect is a physical phenomenon that occurs in the quantum vacuum, the lowest-energy state of the electromagnetic field. The effect is named after the Dutch physicist Hendrik Casimir, who first predicted it in 1948.

The Casimir Effect occurs when two parallel uncharged metallic plates are placed in a vacuum very close to each other, so that the distance between the plates is much smaller than the wavelength of the electromagnetic waves in the vacuum. The vacuum between the plates is not truly empty, but is filled with fluctuating electromagnetic fields, including virtual particles that are constantly popping in and out of existence.

The presence of the plates changes the allowed wavelengths of the virtual particles, leading to a decrease in the energy density of the vacuum between the plates compared to the vacuum outside the plates. This creates a force that pushes the plates together, even in the absence of any external forces. The force due to the Casimir Effect is very small, but it has been observed experimentally using precision equipment.

The Casimir force per unit area F_c/A for idealized, perfectly conducting plates with vacuum between them is

$$\frac{F_c}{A} = -\frac{d}{da}\frac{\langle E \rangle}{A} = -\frac{\hbar c \pi^2}{240 a^4}$$

where, $\hbar$ (hbar) is the reduced Planck constant, c is the speed of light, a is the distance between the plates. The force is negative (attractive): by moving the two plates closer together, the energy is lowered.

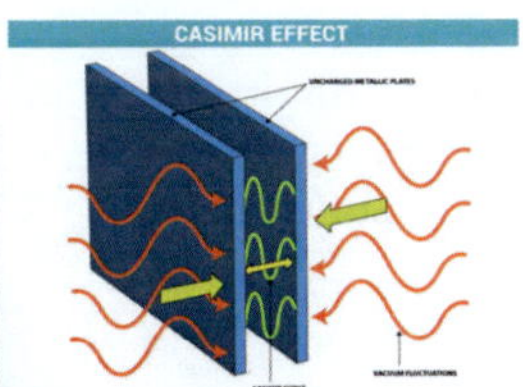

Hendrik Casimir

(Credit:https://commons.wikimedia.org/wiki/File:Hendrik_Casimir_%281958%29.jpg)

Applications of Casimir Effect

Casimir Effect is a fascinating phenomenon that arises from the quantum nature of the vacuum and has important implications for both experimental and theoretical physics. The Casimir Effect has important implications for nanotechnology and has been proposed as a way to create small-scale mechanical devices that could be used in microelectronics and other fields. The Casimir Effect has several potential applications in various fields of science and technology. Here are a few examples:

- **Nanotechnology:** One potential application of the Casimir Effect is in nanotechnology. The force generated by the Casimir Effect can be used to manipulate and control the motion of nanoscale objects, such as nanoparticles and nanotubes. This has potential applications in the development of small-scale mechanical devices and sensors.
- **Microelectromechanical systems (MEMS):** The Casimir Effect has been proposed as a way to create frictionless microscale devices that could be used in MEMS. These devices could have a wide range of applications, including in microfluidics, micro-robotics, and micro-electromechanical sensors.
- **Energy harvesting:** The Casimir Effect can also be used to generate energy. By controlling the distance between two plates, researchers have shown that it is possible to create a fluctuating electric field that can be used to generate a small amount of electrical energy. This has potential applications in the development of energy harvesting devices that could be used to power small-scale electronic devices.
- **Fundamental physics:** The Casimir Effect has also been used in experiments to test fundamental physics theories, such as quantum field theory and the existence of extra dimensions. By measuring the strength of the Casimir Effect under different conditions, researchers can test these theories and gain insight into the nature of the universe.

The Casimir Effect is also important in theoretical physics, as it provides evidence for the existence of virtual particles and the quantum nature of the electromagnetic field. The effect has been studied in the context of various theories of quantum gravity, including string theory and loop quantum gravity.

Casimir Effect has a wide range of potential applications in various fields of science and technology, from nanotechnology and MEMS to energy harvesting and fundamental physics. While these applications are still in the early stages of development, they have the potential to lead to significant advances in these fields.

23. Chandrasekhar Limit

The Chandrasekhar limit is the maximum mass that a white dwarf star can have before it collapses under its own gravitational forces and explodes as a supernova. This limit is named after the Indian astrophysicist Subrahmanyan Chandrasekhar, who first calculated it in 1931.

The Chandrasekhar limit is based on the properties of electron degeneracy pressure, which is the pressure exerted by the electrons in a star's core. This pressure resists the gravitational forces that try to collapse the star. However, as the mass of the star increases, the gravitational forces become stronger, and eventually the electron degeneracy pressure is not enough to support the star against collapse. The Chandrasekhar limit is the maximum mass at which electron degeneracy pressure can support a star against collapse.

The value of the Chandrasekhar limit is approximately 1.4 times the mass of the Sun (1.4 solar mass). Stars that are more massive than this will eventually undergo a core-collapse supernova, in which the core of the star collapses into a neutron star or black hole, while the outer layers of the star are expelled in a massive explosion. The Chandrasekhar limit equation is:

$$M_{limit} = \frac{\omega_3^0\sqrt{3\pi}}{2}\left(\frac{\hbar c}{G}\right)^{\frac{3}{2}}\frac{1}{(\mu_0 m_H)^2}$$

where, ħ refers to the reduced Planck constant. c refers to the speed of light. G denotes the gravitational constant. μ_0 is the average molecular weight. m_H refers to the hydrogen atom mass. $\omega^0_3 \approx 2.018236$ is a constant.

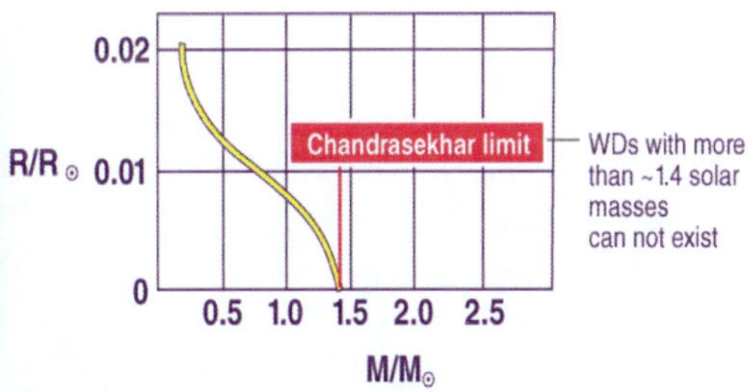

Subrahmanyan Chandrasekhar
(Credit:https://starchild.gsfc.nasa.gov/docs/StarChild/whos_who_level1/chandra.html)

Applications of Chandrasekhar Limit

The Chandrasekhar limit has important implications for the evolution of stars and the universe as a whole. It determines the fate of many types of stars, and it also sets an upper limit on the mass of white dwarf stars, which are important objects in astronomy and astrophysics. The study of the Chandrasekhar limit has led to many important discoveries and insights into the behavior of stars and the structure of the universe. Chandrasekhar limit is a fundamental concept in astrophysics that has numerous applications and implications for the study of stars, supernovae, black holes, and the structure and evolution of the universe.

- **White dwarf stars:** The Chandrasekhar limit sets the maximum mass that a white dwarf star can have. White dwarf stars are the remnants of low- to medium-mass stars (up to 8 solar masses) that have exhausted their nuclear fuel and have shed their outer layers, leaving behind a hot, dense core supported by electron degeneracy pressure. If the mass of the white dwarf exceeds the Chandrasekhar limit, it will undergo a catastrophic collapse and explosion as a Type Ia supernova.
- **Type Ia supernovae:** Type Ia supernovae are among the most luminous and powerful explosions in the universe. They occur when a white dwarf star in a binary system accretes matter from a companion star until it exceeds the Chandrasekhar limit and undergoes a runaway fusion reaction, causing the star to explode. Because the Chandrasekhar limit is a well-defined mass threshold, Type Ia supernovae have a consistent luminosity and can be used as "standard candles" to measure the distances to faraway galaxies and to study the expansion rate of the universe.
- **Stellar evolution:** The Chandrasekhar limit plays a crucial role in the evolution of stars, particularly in the late stages of their lives. The limit sets a maximum mass for white dwarf stars and also determines the minimum mass required for a star to undergo a core-collapse supernova. By studying the masses and properties of stars, astronomers can use the Chandrasekhar limit to understand how stars evolve and how they contribute to the chemical enrichment of the universe.
- **Black holes:** The Chandrasekhar limit is also relevant to the formation of black holes, which are thought to be the end state of massive stars that have undergone a core-collapse supernova. Stars that have a mass above the Chandrasekhar limit will eventually collapse into a black hole without undergoing a supernova explosion. The study of black holes and their properties is an active area of research in astrophysics, and the Chandrasekhar limit provides a key theoretical framework for understanding their formation and behavior.

24. CHARLES'S LAW

Charles's law, also known as the law of volumes, is an experimental gas law that describes the relationship between the volume and temperature of a gas at constant pressure.

The law states that, at constant pressure, the volume of a given mass of an ideal gas is directly proportional to its absolute temperature. Mathematically, Charles's law can be expressed as:

$$V = k \times T$$

where V is the volume of the gas, T is its absolute temperature (measured in kelvin), and k is a constant of proportionality.

This law was first formulated by the French physicist Jacques Charles in the late 18th century, based on his experiments with gases at different temperatures. Charles's law is related to the kinetic theory of gases, which states that the temperature of a gas is related to the average kinetic energy of its particles.

Charles's law is an important concept in the study of thermodynamics and is widely used in various fields of science and engineering, such as in the design of engines and refrigeration systems. It also helps to explain many everyday phenomena, such as why a balloon expands when it is heated or why the pressure in a container increases when it is heated.

Jacques Charles

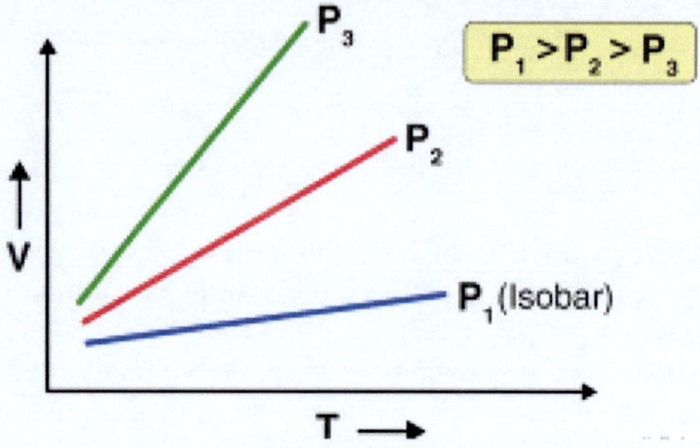

Applications of Charles's Law

Charles's law is an important principle in thermodynamics and has a wide range of practical applications in various fields. Charles's law, also known as the law of volumes, has several important applications, including:

- **Balloon and airship design:** Balloons and airships use hot air to become buoyant, and Charles's law helps explain why. When the air inside the balloon is heated, it expands, causing the volume to increase. Since the density of the air inside the balloon decreases, the balloon becomes lighter and rises.
- **Thermometers:** Charles's law can be used to explain how gas thermometers work. As the temperature of a gas increases, its volume increases, causing the gas to expand and push a column of liquid (such as mercury) up a narrow tube. The height of the liquid column is directly proportional to the temperature of the gas.
- **Gasoline and diesel engines:** Charles's law helps explain why internal combustion engines require a cooling system. When the fuel-air mixture is ignited inside the engine, it expands rapidly and produces a large amount of heat. The cooling system helps regulate the temperature of the engine, preventing it from overheating and causing damage.
- **Industrial processes:** Charles's law is used in many industrial processes that involve gases, such as combustion, fermentation, and refrigeration. It is important to understand the relationship between temperature and volume when designing and operating these processes.

25. Clausius Statement

The Clausius Statement, also known as the Clausius Second Law of Thermodynamics, is a fundamental principle in thermodynamics that relates to the direction of heat flow between two bodies. The statement was formulated by German physicist Rudolf Clausius in the mid-19th century.

The Clausius Statement can be stated as follows: "Heat cannot flow from a colder body to a hotter body without the aid of an external work input."

This statement implies that heat will always spontaneously flow from a hotter body to a colder body, and that in order to reverse this flow, external work must be done. It also suggests that there is a natural tendency for energy to disperse, or for entropy to increase, over time, and that any attempt to create a perpetual motion machine that violates this tendency is impossible. In other words, there are always energy losses due to friction, heat transfer, and other factors that cannot be completely avoided.

The second law has important implications for a wide range of fields, including engineering, chemistry, biology, and environmental science. It sets limits on the efficiency of engines, the design of energy systems, and the sustainability of natural processes.

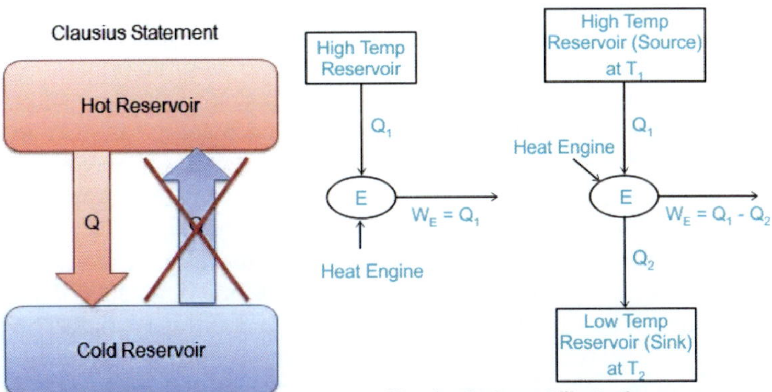

Clausius Statement Illustration

Rudolf Clausius

Applications of Clausius Statement

The Clausius Statement, which is a fundamental principle in thermodynamics, has many practical applications in different fields and its understanding is crucial for engineers and scientists working in energy conversion, thermodynamics, and related fields. Here are some of the applications :

- **Heat engines:** The Clausius Statement is used to explain why heat engines, such as internal combustion engines and steam turbines, are not 100% efficient. It states that in any heat engine cycle, there will always be a net increase in entropy, which means that some energy is always lost to the surroundings.
- **Refrigeration and air conditioning:** The Clausius Statement is also used to explain how refrigeration and air conditioning systems work. These systems use external work to transfer heat from a colder body to a hotter body, which is only possible because of the aid of an external work.
- **Power generation:** The Clausius Statement is applied in the design of power plants, such as nuclear and thermal power plants. It helps engineers to understand the efficiency of the system and the amount of heat energy that is lost to the environment during the generation of electricity.
- **Chemical reactions:** The Clausius Statement is used to understand the direction of chemical reactions and the feasibility of a reaction. It helps chemists to predict the entropy change of a reaction and whether it will occur spontaneously or require an external energy input.
- **Biology:** The law has important implications for the behavior of biological systems, such as living organisms and ecosystems. It sets limits on the efficiency of metabolic processes and the transfer of energy between organisms in food chains.

26. CONTINUITY EQUATION

The continuity equation is a fundamental principle in physics that states that the rate of flow of a fluid or gas must remain constant in a closed system. It is a statement of the conservation of mass, and it applies to both incompressible and compressible fluids. The continuity equation is given by:

$$\nabla \cdot \mathbf{J} = -\partial\rho/\partial t$$

where $\nabla \cdot J$ is the divergence of the flow velocity vector, J, and ρ is the density of the fluid. This equation implies that the rate of change of mass within a given volume of the fluid is equal to the difference between the inflow and outflow rates.

Leonhard Euler

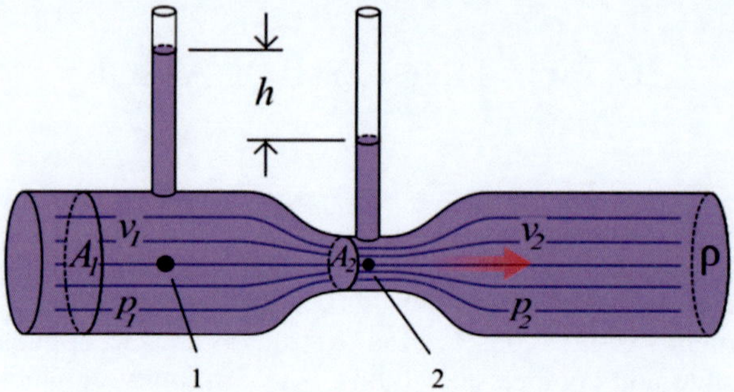

Applications of Continuity Equation

The continuity equation is a fundamental principle that underpins many important concepts in physics, engineering, and other fields. Its applications are diverse and far-reaching, and it remains an essential tool for understanding the behavior of fluids, gases, and other types of systems in motion. Some of its key applications include:

- **Fluid dynamics:** The continuity equation is widely used in fluid dynamics to describe the behavior of fluids in motion. For example, it can be used to analyze the flow of fluids through pipes, nozzles, and other types of conduits.
- **Conservation laws:** The continuity equation is closely related to the conservation laws of physics, such as the conservation of mass and energy. It is a powerful tool for understanding how these laws apply to different types of systems and processes.
- **Atmospheric sciences:** The continuity equation is used extensively in atmospheric sciences to model the flow of air and other gases in the atmosphere. It is particularly useful for predicting the behavior of weather systems and atmospheric pollutants.
- **Electrical engineering:** The continuity equation can also be applied to the flow of electrical current in conductive materials. It is an essential principle for understanding the behavior of circuits and other electrical systems.

27. Coulomb's Law

Coulomb's law is a fundamental principle in physics that describes the electrostatic interaction between charged particles. It states that the force between two charged particles is directly proportional to the product of their charges and inversely proportional to the square of the distance between them. Mathematically, Coulomb's law can be expressed as:

$$F = k \times q_1 \times q_2 / r^2$$

where F is the electrostatic force between the two particles, q_1 and q_2 are their charges, r is the distance between them, and k is the Coulomb constant.

Charles-Augustin de Coulomb

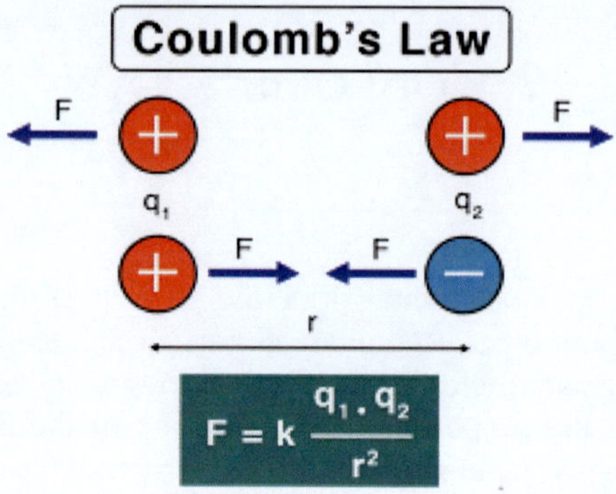

Applications of Coulomb's Law

Coulomb's law is a fundamental principle in physics that has numerous applications in fields ranging from atomic and molecular interactions to electrical circuits and plasma physics. Applications of Coulomb's law include:

♦ **Understanding atomic and molecular interactions:** Coulomb's law is used to understand the interactions between charged particles, such as electrons and protons in atoms and molecules. The force of attraction between oppositely charged particles and the force of repulsion between like charges are described by Coulomb's law.

♦ **Electrostatics:** Coulomb's law is the basis for the study of electrostatics, which involves the behavior of stationary electric charges. It is used to calculate the electrostatic force between two charged objects and to determine the distribution of charges on conductors.

♦ **Electrical circuits:** Coulomb's law is used in the study of electrical circuits, where it helps to calculate the force between charges moving in a wire and to determine the capacitance of a capacitor.

♦ **Particle accelerators:** Coulomb's law is used in the design and operation of particle accelerators, which accelerate charged particles to high speeds. Coulomb's law is used to calculate the force on a charged particle in an electric field, which is used to guide and accelerate the particles.

♦ **Plasma physics:** Coulomb's law is used to understand the behavior of plasmas, which are ionized gases that contain charged particles. It is used to calculate the forces between charged particles in plasmas and to determine the behavior of plasmas in electric and magnetic fields.

28. CURIE'S LAW

Curie's Law is a fundamental principle in physics that describes the magnetic properties of materials. It was named after the physicist Pierre Curie, who first proposed the relationship between magnetization and temperature in 1895.

Curie's Law states that the magnetization of a material is directly proportional to the strength of the magnetic field and the inverse of the absolute temperature. This relationship is expressed mathematically as:

$$M = C / T$$

where M is the magnetization, C is a constant, and T is the absolute temperature. According to Curie's Law, as the temperature of a material increases, its magnetization decreases, and it becomes less magnetic.

Curie's Law is important for understanding the behavior of magnetic materials, such as ferromagnets, paramagnets, and diamagnets. Ferromagnetic materials are strongly magnetic and are used in many applications, including data storage, motors, and generators. Paramagnetic materials are weakly magnetic and are used in magnetic resonance imaging (MRI) machines and other medical equipment. Diamagnetic materials are non-magnetic and are used in many everyday objects, such as copper wires and aluminum foil.

Pierre Curie

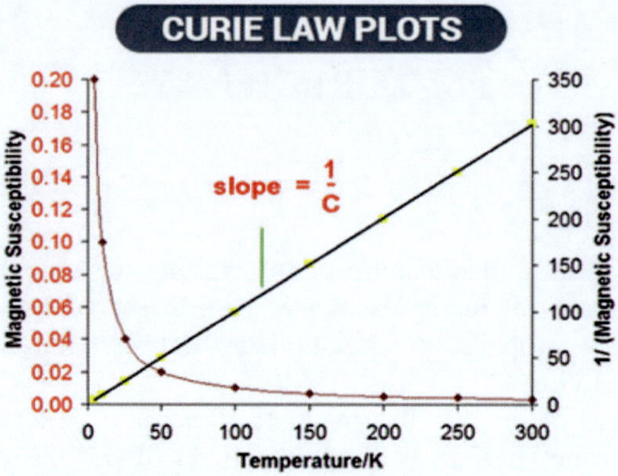

Applications of Curie's Law

Curie's Law is an important principle in physics, which has many practical applications in materials science, engineering, and medical imaging. Its applications are diverse and far-reaching, from designing magnetic storage devices to studying the magnetic properties of the Earth's crust. Some of the applications of Curie's Law are:

- **Magnetic storage:** Curie's Law is used in the design of magnetic storage devices such as hard drives, where magnetic susceptibility plays a crucial role. By understanding the behavior of magnetic materials in response to temperature and magnetic fields, engineers can design more efficient and reliable storage devices.
- **Magnetic resonance imaging (MRI):** In medical imaging, Curie's Law is used to analyze the behavior of magnetic materials in the body, which is important for understanding how they interact with magnetic fields. This information is used to create images of the body's internal structures in MRI scans.
- **Magnetic materials research:** Curie's Law is used in materials research to understand the magnetic properties of materials, such as ferromagnetic and paramagnetic materials. This information is important for developing new materials for use in electronics, energy storage, and other applications.
- **Geophysics:** Curie's Law is used in geophysics to study the magnetic properties of rocks and minerals, which can provide important information about the Earth's magnetic field and the history of the planet's magnetic activity.

29. CURIE-WEISS LAW

The Curie-Weiss Law is a fundamental principle in the study of magnetism, which describes the behavior of a material when it is placed in a magnetic field. The law was first proposed by Pierre Curie and his brother Jacques Curie, and was later refined by Pierre-Ernest Weiss.

The Curie-Weiss Law describes the relationship between the magnetic susceptibility of a material and its temperature. Magnetic susceptibility is a measure of how easily a material can be magnetized, and is defined as the ratio of the magnetic moment to the magnetic field strength. According to the Curie-Weiss Law, the magnetic susceptibility χ of a material is proportional to the inverse of the difference between the temperature and a constant called the Curie temperature (T_c).

The Curie–Weiss law describes the magnetic susceptibility of a ferromagnet in the paramagnetic region above the Curie point:

$$\chi = C / (T - T_C)$$

where C is a material-specific Curie constant, T is the absolute temperature, and T_C is the Curie temperature, both measured in kelvin. The law predicts a singularity in the susceptibility at $T = T_C$. Below this temperature, the ferromagnet has a spontaneous magnetization.

The Curie temperature is the temperature at which a ferromagnetic material undergoes a phase transition and loses its magnetic properties. Below the Curie temperature, the magnetic moments of the material are aligned and the material behaves as a magnet. Above the Curie temperature, the magnetic moments are randomly oriented and the material does not exhibit magnetic properties.

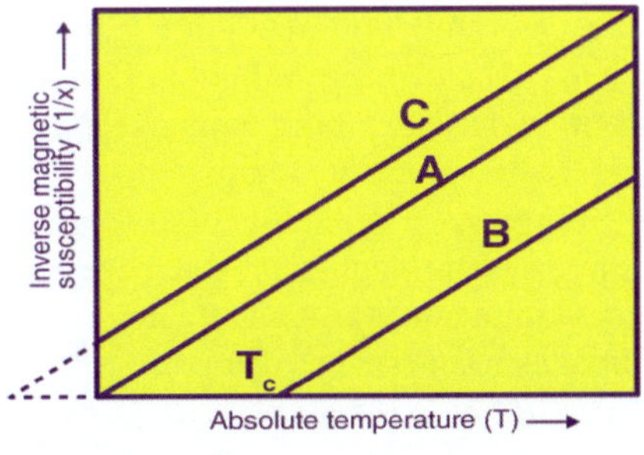

Pierre Curie **Pierre Weiss**
(https://www.nobelprize.org/prizes/physics/1903/pierre-curie/biographical/)

Applications of Curie-Weiss Law

Curie-Weiss law is an important physical law that has many practical applications in materials science, electrical engineering, medical imaging and in the study of magnetism and magnetic materials. Its applications are diverse and far-reaching, from designing magnetic components to producing detailed medical images of the body.

- **Magnetic materials:** The Curie-Weiss law is used to explain the magnetic properties of materials, such as iron, cobalt, and nickel. By understanding the relationship between temperature and magnetic susceptibility, scientists and engineers can design more efficient magnetic materials for use in a variety of applications.
- **Materials science:** The Curie-Weiss law is used in materials science to study the magnetic properties of materials, including ferromagnetic and paramagnetic materials. By analyzing the temperature-dependent magnetic susceptibility of materials, researchers can determine the behavior of these materials at different temperatures.
- **Electrical engineering:** The Curie-Weiss law is used in electrical engineering to design and optimize the performance of magnetic components, such as transformers and inductors. By understanding the behavior of magnetic materials at high temperatures, engineers can design more efficient and reliable magnetic components.
- **Medical imaging:** The Curie-Weiss law is used in medical imaging technologies, such as magnetic resonance imaging (MRI). By using magnetic fields to manipulate the spin of atoms in the body, MRI can produce detailed images of internal structures.

30. D'ALEMBERT'S PRINCIPLE

D'Alembert's principle is a fundamental principle in classical mechanics named after French mathematician Jean le Rond d'Alembert. It states that the motion of a system of particles can be described by the net force acting on the system and the mass of the system.

Specifically, D'Alembert's principle states that the sum of the forces acting on a system of particles, when augmented by a term proportional to the acceleration of the particles and in the opposite direction, is equal to zero. Mathematically, this can be written as:

$$\Sigma F + (-m) a = 0$$

where ΣF is the sum of the external forces acting on the system, m is the mass of the system, and a is the acceleration of the system.

This principle is often used to simplify problems in mechanics, as it allows the equations of motion to be written in terms of the net force acting on a system, rather than considering the individual forces acting on each particle in the system.

Jean le Rond D'Alembert

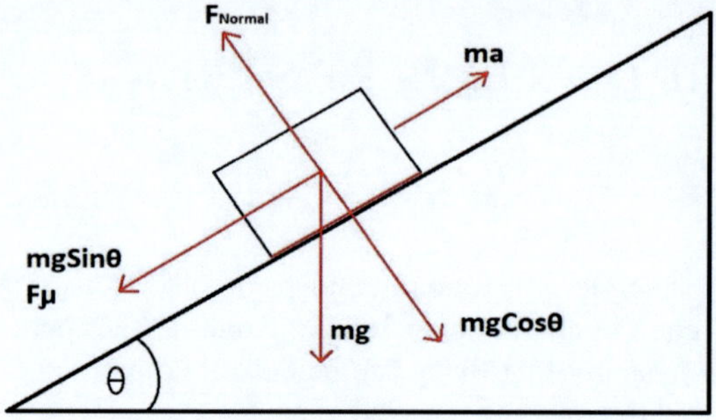

Free Body Diagram of Forces Acting on a Car Accelerating up an Inclined Plane

Applications of D'Alembert's Principle

D'Alembert's principle has numerous applications in classical mechanics:

- **Dynamics of constrained systems:** D'Alembert's principle is used to determine the equations of motion for systems that are subject to constraints, such as a pendulum swinging on a fixed pivot or a car moving along a curved road. The principle can be used to reduce the number of degrees of freedom in the system and simplify the analysis.
- **Lagrange's equations:** D'Alembert's principle is one of the foundations of Lagrangian mechanics, which is a powerful method for describing the motion of systems with many degrees of freedom. By using the principle to eliminate forces of constraint, one can derive Lagrange's equations, which provide a concise and elegant way to describe the motion of a system.
- **Virtual work:** D'Alembert's principle can be used to derive the principle of virtual work, which is a useful tool in engineering and physics. The principle of virtual work states that the work done by a system of forces is equal to the change in the potential energy of the system, and it is used to analyze the stability of structures and the behavior of machines.
- **Analyzing mechanical systems:** D'Alembert's principle is used to analyze the motion of mechanical systems, such as machines and vehicles. By applying the principle, engineers can determine the forces acting on the system and design it to operate more efficiently and safely.

D'Alembert's principle is a fundamental concept in mechanics that is used to simplify complex problems and provide insight into the behavior of mechanical systems.

31. Dalton's Law of Multiple Proportions

Dalton's law of multiple proportions states that when two elements combine to form a series of compounds, the ratios of the masses of the second element that combine with a fixed mass of the first element can be expressed as ratios of small whole numbers. This law was proposed by the English chemist John Dalton in the early 19th century and is considered one of the fundamental laws of chemistry. The law helped to explain why some elements can combine in different proportions to form different compounds.

John Dalton

Nitrogen Monoxide (NO)	Nitrogen Dioxide (NO_2)
mass of N: mass of O	mass of N: mass of O
↓	↓
14:16	14:32
↓	↓
$^{14}/_{14}:{}^{16}/_{14}$	$^{14}/_{14}:{}^{32}/_{14}$
↓	↓
1:1.143	1:2.286

Applications of Dalton's Law of Multiple Proportions

Dalton's law of multiple proportions is a fundamental concept in chemistry and has a wide range of applications. Some of the applications of this law include:

- **Determining the composition of compounds:** Dalton's law of multiple proportions states that when two elements form multiple compounds, the ratio of the masses of one element that combine with a fixed mass of the other element is always in simple whole-number ratios. This principle is widely used in chemistry to determine the composition of compounds.
- **Stoichiometry:** The law of multiple proportions is important in stoichiometry, the calculation of the quantities of reactants and products in a chemical reaction. The law allows chemists to determine the exact proportions of reactants needed to produce a given amount of product.
- **Isotope analysis:** The law of multiple proportions can be used in isotope analysis to determine the isotopic composition of elements in a sample. The ratios of different isotopes of an element in a sample can be compared to the ratios predicted by the law of multiple proportions to determine the isotopic composition.
- **Analyzing combustion reactions:** Combustion reactions, in which a fuel is burned in the presence of oxygen to produce carbon dioxide and water, are important in many industrial and environmental applications. The law of multiple proportions can be used to determine the amounts of fuel and oxygen needed to produce a given amount of carbon dioxide and water.
- **Identifying unknown compounds:** The law of multiple proportions can be used to identify unknown compounds. By analyzing the ratios of the masses of the elements in the compound, chemists can determine the likely chemical formula of the compound.

32. Darcy's Law

Darcy's law is a fundamental principle of fluid mechanics that describes the flow of fluids through porous media, such as soils, rocks, and groundwater systems.

The law states that the rate of fluid flow through a porous medium is directly proportional to the pressure gradient driving the flow and to the permeability of the medium. Mathematically, Darcy's law can be expressed as:

$$Q = -K A (dh/dl)$$

where Q is the flow rate of fluid through the medium, K is the permeability of the medium (a measure of its ability to transmit fluids), A is the cross-sectional area of the medium through which the fluid is flowing, dh/dl is the hydraulic gradient (the change in pressure per unit distance), and the negative sign indicates that the flow is in the direction of decreasing pressure.

Darcy's law was first formulated by the French engineer Henry Darcy in 1856, based on his experiments with water flow through sand columns. The law is widely used in hydrogeology, petroleum engineering, and soil mechanics, among other fields, to model and predict the flow of fluids through porous media. It is also used in the design and management of groundwater resources, such as in the construction of wells and groundwater remediation systems.

Henry Darcy

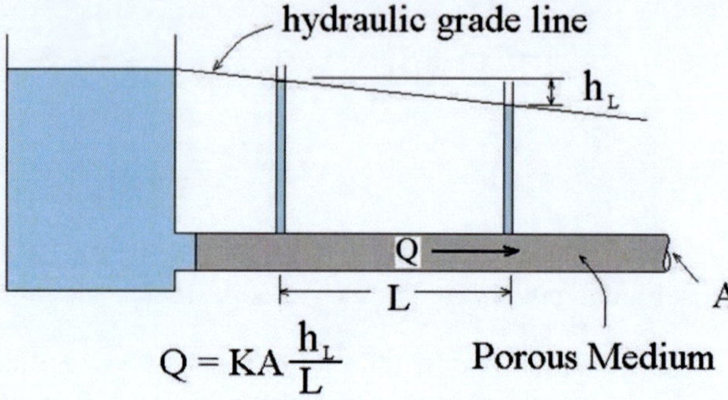

Darcy's Law Apparatus

Applications of Darcy's Law

Darcy's law is a fundamental principle in hydrogeology and is widely used to describe groundwater flow through porous media. Some applications of Darcy's law are:

- **Groundwater modeling:** Darcy's law is used to develop mathematical models that can simulate groundwater flow and predict the behavior of an aquifer system.
- **Contaminant transport:** Darcy's law is used to model the movement of contaminants in groundwater systems. This is important in assessing the potential risks associated with contaminated sites.
- **Water resource management:** Darcy's law is used to assess the availability of groundwater resources in a given area. This is important in determining the sustainability of groundwater use and the potential impacts of groundwater withdrawals.
- **Geotechnical engineering:** Darcy's law is used to model the movement of water through soils and rock formations. This is important in designing structures such as dams, levees, and tunnels.
- **Petroleum engineering:** Darcy's law is used to model the flow of fluids through porous rock formations in oil and gas reservoirs. This is important in predicting the behavior of these reservoirs and optimizing production.

33. De Broglie Wavelength

De Broglie wavelength is a fundamental concept in quantum mechanics that describes the wave-like behavior of matter particles, such as electrons, protons, and atoms.

The concept was introduced by the French physicist Louis de Broglie in 1924, who proposed that if electromagnetic radiation, such as light, can exhibit both wave-like and particle-like behavior, then perhaps matter particles could also exhibit wave-like behavior. De Broglie suggested that any particle, including an electron or a proton, has a wavelength associated with it, which is related to its momentum.

The de Broglie wavelength (λ) of a particle is given by the following equation:

$$\lambda = h / p$$

where h is Planck's constant (a fundamental constant of nature) and p is the momentum of the particle. This equation shows that the de Broglie wavelength of a particle decreases as its momentum increases.

The de Broglie wavelength has important implications in quantum mechanics, as it predicts that matter particles can exhibit wave interference and diffraction, just like electromagnetic waves. This wave-like behavior is central to many quantum mechanical phenomena, such as the wave-particle duality of matter and the formation of electron interference patterns in diffraction experiments. The de Broglie wavelength is also used in various fields of science and technology, such as in electron microscopy and in the design of quantum devices.

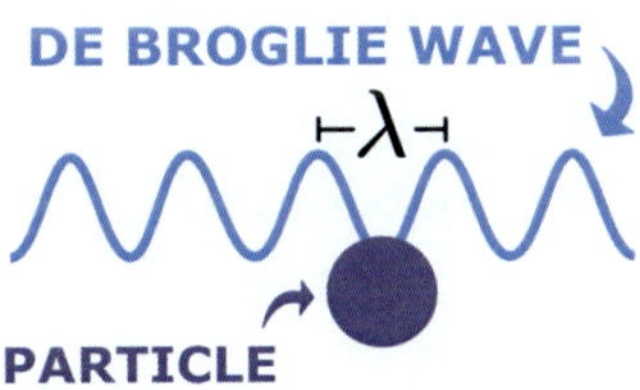

$$\lambda = \frac{h}{mv} = \frac{h}{p}$$

Louis de Broglie

Applications of De Broglie Wavelength

De Broglie wavelength has many applications in various fields, including physics, chemistry, and material science. De Broglie wavelength is a concept in quantum mechanics that is used to describe the wave-like behavior of particles, particularly in the context of the dual nature of matter. Here are some applications of De Broglie wavelength:

- **Diffraction:** De Broglie wavelength is used to describe the diffraction of particles, such as electrons, through a crystal lattice. This phenomenon is used in various applications, including X-ray crystallography, which is used to determine the structure of molecules.
- **Electron microscopy:** De Broglie wavelength is also used in electron microscopy. In this technique, the De Broglie wavelength of electrons is used to image the structure of materials at the atomic scale.
- **Particle accelerators:** De Broglie wavelength is also used in particle accelerators, where particles are accelerated to high energies and then made to collide with a target. The wavelength of the particles is used to determine the size of the colliding particles and the energy needed to produce certain reactions.
- **Quantum computing:** De Broglie wavelength is also used in the development of quantum computers. The wavelength of particles is used to control and manipulate the quantum states of particles in a quantum computer.
- **Bose-Einstein condensates:** De Broglie wavelength is also used to describe the behavior of Bose-Einstein condensates. In these condensates, particles behave as waves, and their De Broglie wavelength determines the size of the condensate and the way the particles interact with each other.

34. Dermott's Law

Dermott's law, also known as Dermott's rule, is an astronomical principle that describes the relationship between the periods of natural satellites (moons) in a planetary ring system. It states that the radial drift of body in a planetary ring is proportional to the square of their size. This means that smaller particles experience more significant radial drift than larger particles.

The equation for Dermott's law is given by:

$$dr/dt = -C \times r^2$$

where dr/dt is the rate of change of the radial distance of a particle from the planet, r is the radial distance, and C is a constant that depends on the properties of the planetary ring system, such as the mass of the planet, the density of the particles, and the viscosity of the surrounding medium.

Dermott's law has important implications for the structure and evolution of planetary ring systems, as it can lead to the segregation of particles by size and the formation of gaps and sharp edges in the ring. It also helps explain why the innermost and outermost regions of planetary rings are often dominated by particles of different sizes. Dermott's law was first proposed by the American astronomer Stanley Dermott in 1972, based on his studies of the orbital dynamics of comets. The law suggests that comets with longer periods tend to have smaller inclinations to the ecliptic plane, while comets with shorter periods tend to have larger inclinations. This is because comets with larger inclinations tend to experience stronger perturbations from the gravitational interactions with the planets, which can cause their orbits to become unstable and either collide with a planet or be ejected from the Solar System.

Dermott's law has been used to study the distribution and evolution of comets in the Solar System, as well as to predict the orbits of newly discovered comets. It has also been applied in the study of exoplanetary systems, where it can help to infer the presence and properties of unseen planets based on the observed orbital parameters of the system's comets.

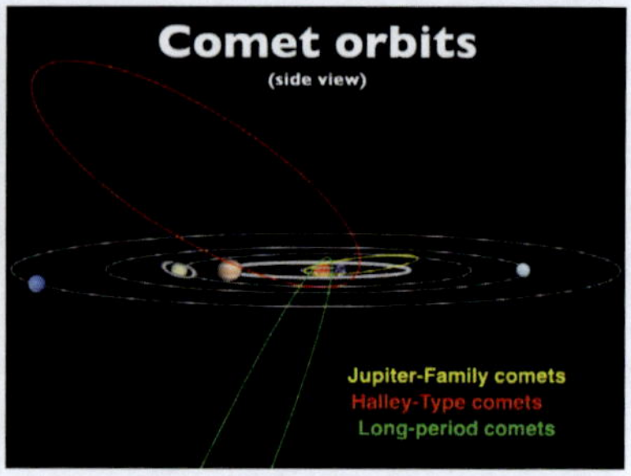

Stanley Dermott
(Credit: https://astro.ufl.edu/directory/stanley-dermott/)

Applications of Dermott's Law

Dermott's law, also known as the "lost comet" law, describes the tendency of short-period comets to be ejected from the solar system due to gravitational interactions with the planets. It has applications in astronomy, planetary science, and planetary defense. Here are some applications of Dermott's law:

♦ **Understanding comet populations:** Dermott's law helps us understand the population of short-period comets in the solar system. These comets have periods of less than 200 years and are believed to originate from the Kuiper Belt and the scattered disk. Dermott's law explains why the number of short-period comets is much smaller than the number of long-period comets, which are not affected by the gravitational interactions of the planets.

♦ **Identifying lost comets:** Dermott's law can be used to identify lost comets, which are comets that have been ejected from the solar system and are no longer observable. By studying the orbits of known short-period comets, astronomers can predict the orbits of lost comets and search for them using telescopes.

♦ **Modeling solar system dynamics:** Dermott's law is an important factor in models of solar system dynamics. These models help us understand the evolution of the solar system over time and can be used to make predictions about the future behavior of the planets and comets.

♦ **Planetary defense:** Understanding the behavior of comets is important for planetary defense. Comets have the potential to collide with Earth and cause catastrophic damage. By studying the orbits of comets and predicting their behavior, astronomers can identify potentially hazardous objects and take measures to prevent collisions.

35. DIRAC EQUATION

The Dirac equation is a relativistic quantum mechanical equation that describes the behavior of spin-1/2 particles, such as electrons, in a relativistic quantum field theory. The equation was developed by British physicist Paul Dirac in 1928 and is an important part of the standard model of particle physics.

$$i\hbar\frac{\partial}{\partial_t}\psi(x,t) = \left(\beta mc^2 + c\left(\sum_{n=1}^{3}\alpha_n p_n\right)\right)\psi(x,t)$$

where,
$\psi=\psi(x,t)$ is the electron wave function
m is the electron mass at rest
x, t is the space time coordinates
p1, p2, p3 are the momentum components
c is the speed of light
$\hbar$ is the Planck constant

The Dirac equation is a first-order partial differential equation that combines quantum mechanics and special relativity. It is written in terms of a four-component spinor wave function that describes the electron's quantum state. The Dirac equation predicts that spin-1/2 particles have both positive and negative energy solutions, which led to the discovery of antimatter.

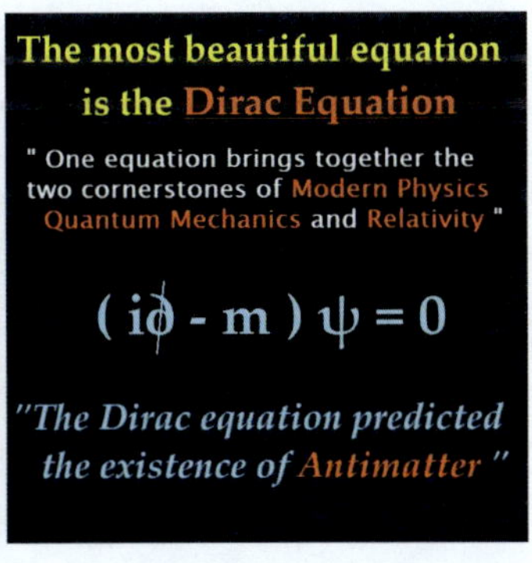

Paul Dirac

Applications of Dirac Equation

The Dirac equation has a wide range of applications in various fields of physics, including particle physics, quantum field theory, condensed matter physics, and cosmology. The equation has many applications in theoretical physics, including the prediction of the existence of antiparticles, the development of quantum electrodynamics (QED), the study of the behavior of electrons in strong magnetic fields, and the theory of superconductivity. The Dirac equation also played a crucial role in the development of quantum field theory and the formulation of the standard model of particle physics.

It is used to study the behavior of electrons in materials with strong spin-orbit coupling and to predict the existence of exotic states of matter, such as topological insulators and topological superconductors.

- **Prediction of antimatter:** One of the most important applications of the Dirac equation is the prediction of the existence of antimatter. The equation predicts the existence of both positive and negative energy solutions, which led Dirac to propose the existence of particles with negative energy, or antiparticles. This prediction was confirmed experimentally with the discovery of the positron, the antiparticle of the electron.
- **Quantum field theory:** The Dirac equation plays a crucial role in the development of quantum field theory, which is the framework for describing the behavior of particles and fields in a relativistic quantum mechanical context. The equation is used to describe the behavior of spin-1/2 particles, such as electrons, in a relativistic quantum field theory.
- **High-energy physics:** The Dirac equation is used to describe the behavior of particles in high-energy physics experiments, such as those at particle accelerators. The equation is used to calculate the scattering amplitudes for various processes, such as electron-positron annihilation and quark-gluon interactions.
- **Condensed matter physics:** The Dirac equation has important applications in condensed matter physics, where it is used to study the behavior of electrons in materials with strong spin-orbit coupling, such as topological insulators and topological superconductors.
- **Cosmology:** The Dirac equation is used in cosmology to study the behavior of particles in the early universe. The equation is used to calculate the abundance of particles, such as neutrinos and dark matter, and their interactions in the early universe.

36. DOPPLER EFFECT

The Doppler Effect is a physical phenomenon that describes the change in frequency of a wave (such as sound or light) when there is relative motion between the source of the wave and an observer. This effect is named after the Austrian physicist Christian Doppler, who first described it in 1842.

The Doppler Effect occurs because waves emitted by a moving source appear to be compressed or stretched depending on the direction of motion relative to the observer. For example, when a car with a siren moves towards an observer, the sound waves from the siren are compressed and have a higher frequency than they would if the car were stationary. Conversely, when the car moves away from the observer, the sound waves are stretched and have a lower frequency.

The mathematical formula for the Doppler Effect is given by:

$$f' = f\,(v \pm v_r) / (v \pm v_s)$$

where f is the original frequency of the wave, f' is the observed frequency, v is the speed of the wave in the medium, v_s is the speed of the source, and v_r is the speed of the observer relative to the medium.

Christian Doppler

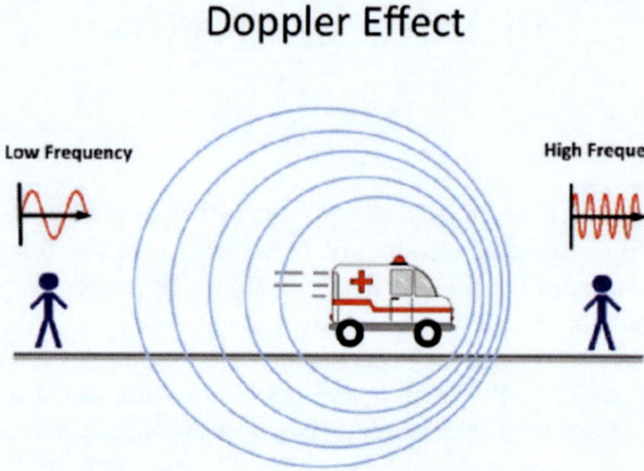

Applications of Doppler Effect

Doppler Effect is an important concept in physics with many practical applications in a wide range of fields, including:

- **Radar and sonar systems:** The Doppler Effect is used in radar and sonar systems to measure the velocity of moving objects. The change in frequency of the reflected waves can be used to calculate the velocity of the object relative to the observer.
- **Medical imaging:** The Doppler Effect is used in medical imaging techniques, such as ultrasound, to measure blood flow and detect abnormalities in the cardiovascular system.
- **Astronomy:** The Doppler Effect is used in astronomy to measure the velocity of stars and galaxies. The change in frequency of light waves emitted by celestial objects can be used to determine whether they are moving towards or away from Earth.
- **Musical instruments:** The Doppler Effect can be used in musical instruments, such as sirens and whistles, to create a change in pitch by moving the source towards or away from the listener.

37. Drake Equation

The Drake equation is a probabilistic argument used to estimate the number of detectable extraterrestrial civilizations in the Milky Way galaxy. The equation was developed by American astronomer Frank Drake in 1961 and is named after him.

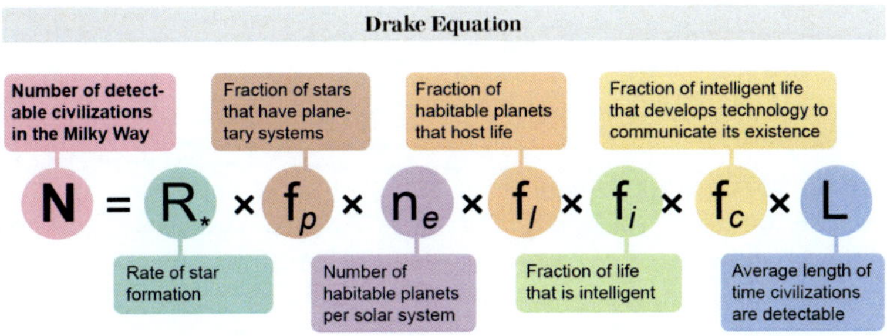

The Drake equation consists of a series of factors that are multiplied together to give an estimate of the number of detectable civilizations. The factors include:

The rate of star formation in the galaxy, The fraction of stars that have planets, The number of habitable planets per star system, The fraction of habitable planets where life emerges, The fraction of life-bearing planets where intelligent life emerges, The fraction of intelligent civilizations that develop technology capable of communicating with others, and The length of time that intelligent civilizations release detectable signals into space.

Each factor is a range of values, as they are still not well understood or constrained. The Drake equation is used to stimulate scientific and philosophical discussions about the search for extraterrestrial life and the factors that might contribute to the emergence and survival of civilizations in the universe.

The Drake equation is not a formal theory, but rather a heuristic tool that helps scientists to frame their discussions and research efforts about the possibility of life beyond Earth. It has been used as a starting point for many studies and debates about astrobiology, the Fermi paradox, and the search for extraterrestrial intelligence.

Frank Drake
(Credit:https://commons.wikimedia.org/wiki/File:Frank_Drake_at_Cornell,_October_2017.jpg)
(Credit: https://pixabay.com/illustrations/space-ship-alien-sci-fi-6937020/)

Applications of Drake Equation

Drake equation has applications in astronomy and astrobiology.

- **Estimating the likelihood of extraterrestrial life:** The Drake equation can be used to estimate the probability of the existence of extraterrestrial life in the universe. By considering the various factors that could influence the development of intelligent life, such as the rate of star formation and the likelihood of habitable planets, the equation provides a way to estimate the probability of intelligent life arising elsewhere in the universe.
- **Searching for extraterrestrial intelligence:** The Drake equation can also be used to guide the search for extraterrestrial intelligence (SETI). By estimating the number of active, communicative extraterrestrial civilizations in the galaxy, astronomers can focus their search efforts on regions of the galaxy that are more likely to contain such civilizations.
- **Understanding the evolution of life:** The Drake equation can be used to help understand the evolution of life on Earth and other planets. By considering the factors that could influence the development of intelligent life, such as the availability of resources and the stability of planetary environments, the equation provides a framework for studying the conditions that are necessary for the evolution of life.
- **Inspiring scientific curiosity:** The Drake equation has inspired scientific curiosity and imagination about the possibility of extraterrestrial life. It has encouraged scientists and the public to think critically about the possibility of life beyond Earth and to explore the universe in search of answers.

38. DULONG-PETIT LAW

Dulong–Petit law, also known as the Law of Dulong and Petit, states that the molar heat capacity of a crystal is approximately constant and equal to 3R, where R is the gas constant. This law was first proposed by French chemists Pierre Louis Dulong and Alexis Thérèse Petit in 1819. The law applies to a wide range of solid elements and compounds at room temperature. The molar heat capacity is the amount of heat needed to raise the temperature of one mole of a substance by one degree Celsius. The law provides a way to calculate the atomic weight of an element, as the product of its heat capacity and the gas constant is approximately equal to its atomic weight. However, the law is not applicable to all solids, and deviations from it have been observed in many cases.

Pierre Louis Dulong

Alexis Thérèse Petit

Energy per mole = $3kTN_A$

k= Boltzmann's constant
T= Temperature in Kelvins
N_A = Avogadro's number

The Law of Dulong & Petit

$$C_v = \frac{\partial}{\partial T}(3kTN_A) = 3kN_A\ /\ mole = 24.94J\ /\ mole\ K$$

Applications of Dulong-Petit Law

The Dulong–Petit law states that the molar specific heat capacity of most solid elements is approximately equal to a constant value of 3R, where R is the gas constant. This law has several applications in the field of materials science and thermodynamics:

- **Predicting melting points:** The Dulong–Petit law can be used to estimate the melting points of metals. According to the law, the heat capacity of metals is directly proportional to the melting point. Therefore, metals with a high heat capacity will have a high melting point.
- **Determining the atomic weight of elements:** The Dulong–Petit law can be used to determine the atomic weight of elements. The law states that the product of the atomic weight and the molar specific heat capacity of an element is constant. Therefore, if the specific heat capacity of an element is known, its atomic weight can be calculated.
- **Understanding heat transfer:** The Dulong–Petit law can help in understanding the transfer of heat in materials. The law predicts that the specific heat capacity of a material is directly proportional to the atomic weight. This means that materials with heavy atoms will have a higher specific heat capacity and therefore transfer heat more slowly than materials with lighter atoms.
- **Predicting the heat of formation of compounds:** The Dulong–Petit law can be used to predict the heat of formation of compounds. The law states that the heat capacity of a compound is directly proportional to the number of atoms in the compound. Therefore, the heat of formation of a compound can be estimated based on the heat capacities of its constituent atoms.

39. EHRENFEST'S THEOREM

Ehrenfest's theorem is a fundamental theorem in quantum mechanics that relates the time evolution of an expectation value of an observable to the time derivative of the corresponding quantum mechanical operator. The theorem is named after the Austrian physicist Paul Ehrenfest, who first formulated it in 1927.

Mathematically, Ehrenfest's theorem can be expressed as follows:

$$d/dt \langle A \rangle = (1/i\hbar) \langle [A,H] \rangle + \langle \partial A/\partial t \rangle$$

where $\langle A \rangle$ denotes the expectation value of the operator A, H is the Hamiltonian of the system, and [A,H] is the commutator of A and H. The operator $\partial A/\partial t$ is the time derivative of A.

Ehrenfest's theorem shows that the time derivative of the expectation value of an observable can be expressed in terms of the commutator of the observable with the Hamiltonian and the time derivative of the observable itself. The commutator represents the degree to which the observable and the Hamiltonian do not commute, or in other words, how their values change with respect to each other over time.

Ehrenfest's theorem has important implications in quantum mechanics, as it relates the classical and quantum descriptions of physical systems. The theorem shows that in the limit of large quantum numbers, quantum mechanics reproduces the classical equations of motion. The theorem is also used in many areas of physics, such as in the study of molecular dynamics, quantum computing, and the behavior of quantum systems in non-equilibrium states.

Paul Ehrenfest

Generalized Ehrenfest Theorem

Quantum Mechanics

$$\frac{d}{dt}\langle Q \rangle = \frac{i}{\hbar}\langle [H, Q] \rangle + \left\langle \frac{\partial Q}{\partial t} \right\rangle$$

Applications of Ehrenfest's Theorem

Ehrenfest's theorem is a fundamental concept in quantum mechanics that relates the time evolution of expectation values of observables to the corresponding classical equations of motion. Ehrenfest's theorem has many applications in various fields of quantum mechanics, including quantum dynamics, quantum algorithms, quantum control theory, and the study of decoherence. It provides a useful tool for relating quantum behavior to classical dynamics, allowing for a better understanding of the behavior of complex quantum systems.

Here are some applications of Ehrenfest's theorem:

- **Understanding quantum dynamics:** Ehrenfest's theorem helps us understand the behavior of quantum systems by relating their time evolution to classical dynamics. This provides a useful tool for studying the behavior of complex quantum systems, such as atoms, molecules, and solids.
- **Designing quantum algorithms:** Ehrenfest's theorem is also used in the design of quantum algorithms. By using the theorem to relate quantum dynamics to classical dynamics, quantum algorithms can be designed to simulate classical dynamics efficiently, allowing for the study of complex systems in new and innovative ways.
- **Developing quantum control theory:** Ehrenfest's theorem is an important tool in the development of quantum control theory, which aims to manipulate and control quantum systems for various applications, such as quantum computing and quantum sensing. By understanding the time evolution of expectation values of observables, control strategies can be developed to manipulate the behavior of quantum systems.
- **Studying decoherence:** Ehrenfest's theorem can also be used to study decoherence, which is the loss of quantum coherence in a system due to its interaction with the environment. By understanding the time evolution of expectation values of observables, researchers can study the effects of decoherence on quantum systems and develop strategies to mitigate its effects.

40. EINSTEIN FIELD EQUATION

The Einstein Field Equation is a set of ten partial differential equations that describe the relationship between the curvature of space-time and the distribution of matter and energy in the universe. The equation was developed by Albert Einstein as part of his theory of general relativity, which describes gravity as the curvature of space-time caused by the presence of matter and energy.

The Einstein Field Equation can be written as:

$$G_{\mu\nu} = 8\pi T_{\mu\nu}$$

where $G_{\mu\nu}$ is the Einstein tensor, which describes the curvature of space-time, and $T_{\mu\nu}$ is the stress-energy tensor, which describes the distribution of matter and energy. The equation essentially states that the curvature of space-time is proportional to the distribution of matter and energy in the universe.

Albert Einstein
(Credit:https://www.bild.bundesarchiv.de/dba/de/search/?query=Bild+183 -19000-1918)

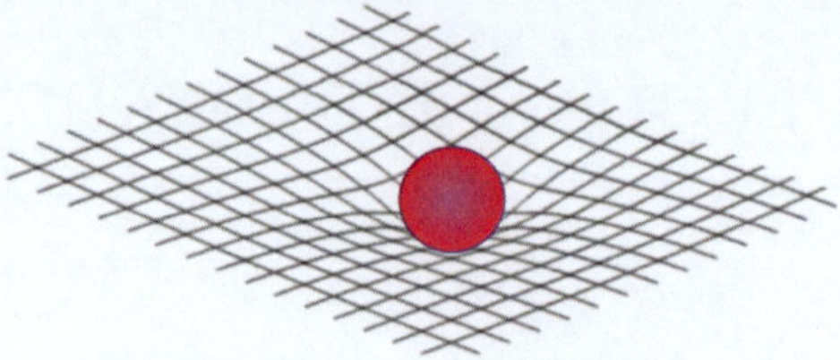

$$G_{\mu\nu}+\Lambda g_{\mu\nu}=\frac{8\pi G}{c^4}T_{\mu\nu}$$

Applications of Einstein Field Equation

The Einstein Field Equation has many implications for our understanding of the universe, including:

- **Black holes:** The Einstein Field Equation predicts the existence of black holes, which are regions of space-time where the curvature is so strong that nothing can escape, not even light.
- **Gravitational waves:** The equation also predicts the existence of gravitational waves, which are ripples in space-time that propagate at the speed of light.
- **Cosmology:** The Einstein Field Equation is used to model the evolution of the universe and to make predictions about its future. It is also used to study the large-scale structure of the universe, such as galaxies and clusters of galaxies.
- **Alternative theories of gravity:** The Einstein Field Equation is a cornerstone of our understanding of gravity, but it is not the only possible theory. There are alternative theories of gravity that modify or extend the equation in various ways, such as modifying the gravitational constant or introducing extra dimensions.

Einstein Field Equation is a fundamental equation in physics that describes the relationship between the curvature of space-time and the distribution of matter and energy in the universe. It has many implications for our understanding of gravity, black holes, gravitational waves, and the structure and evolution of the universe.

41. Einstein's General Theory of Relativity

Einstein's general theory of relativity is a theory of gravity that was developed by Albert Einstein between 1907 and 1915. It is one of the two main pillars of modern theoretical physics, the other being quantum mechanics.

The general theory of relativity is based on two postulates: the principle of equivalence and the principle of general covariance. The principle of equivalence states that the effects of gravity are indistinguishable from the effects of acceleration, while the principle of general covariance states that the laws of physics should be the same in all frames of reference.

Using these postulates, Einstein developed a new theory of gravity that was fundamentally different from the classical theory of gravity developed by Isaac Newton. According to the general theory of relativity, gravity is not a force between masses, but rather a curvature of space-time caused by the presence of mass and energy.

In the general theory of relativity, the curvature of space-time is described by a mathematical object called the metric tensor. The metric tensor determines the distance between two points in space-time, and it is affected by the presence of mass and energy. The motion of particles and objects is then determined by the geometry of space-time.

The general theory of relativity has many important consequences, such as the prediction of the bending of light by gravity, the existence of black holes, and the prediction of gravitational waves. The theory has been extensively tested and confirmed by a wide range of observations, including the bending of starlight during a solar eclipse, the motion of planets in the solar system, and the observations of gravitational waves by the LIGO and Virgo collaborations.

The general theory of relativity is considered one of the most important scientific achievements of the 20th century, and it has had a profound impact on many areas of physics, including cosmology, particle physics, and the study of the structure and evolution of the universe.

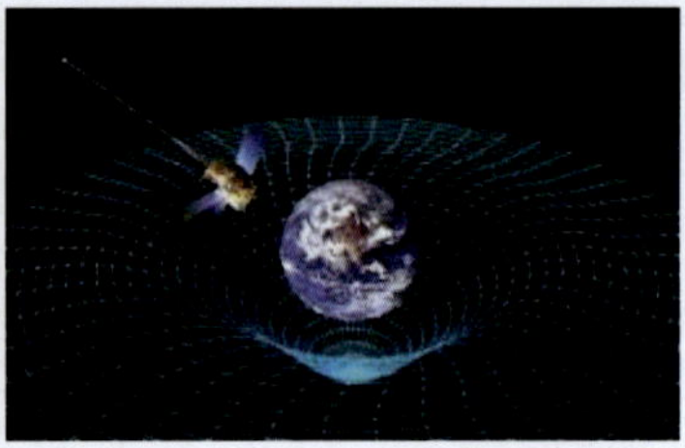

Albert Einstein

Applications of Einstein's General Theory of Relativity

General relativity has many important applications in physics, astronomy, and technology. Here are some applications of the theory:

♦ **Understanding the behavior of black holes:** General relativity predicts the existence of black holes, which are regions of space where the gravitational pull is so strong that nothing, not even light, can escape. The theory provides a way to study the behavior of black holes and the effects of their strong gravitational fields on nearby matter.

♦ **Explaining the behavior of the universe:** General relativity provides a way to study the large-scale structure of the universe and the behavior of its expansion. The theory predicts the existence of dark matter and dark energy, which are thought to make up a significant portion of the universe and influence its expansion.

♦ **Improving the accuracy of GPS:** General relativity predicts that clocks in a strong gravitational field will run slower than clocks in a weaker field. This effect, known as time dilation, has to be taken into account for accurate GPS measurements. Without general relativity, the GPS system would be less accurate by several meters.

♦ **Gravitational wave detection:** General relativity predicts the existence of gravitational waves, which are ripples in the fabric of space-time caused by the acceleration of massive objects. The theory provides a framework for detecting and studying these waves, which were detected for the first time in 2015 by the Laser Interferometer Gravitational-Wave Observatory (LIGO).

♦ **Understanding the behavior of the early universe:** General relativity plays a crucial role in understanding the behavior of the universe in its early stages. The theory predicts the existence of cosmic microwave background radiation, which is thought to be leftover radiation from the Big Bang.

42. Electric Potential

Electric potential is the amount of electrical potential energy per unit charge at a given point in an electric field. It is a scalar quantity that represents the work required to move a unit positive charge from an infinite distance to that point in the electric field. The unit of electric potential is the volt (V).

$$V = k_e \frac{q}{r}$$

where,

V	=	electric potential
k_e	=	Coulomb constant
q	=	charge
r	=	distance of separation

The concept of electric potential is important in many areas of physics, including electromagnetism and electronics.

Alessandro Volta

$$V = \frac{W}{Q} \quad or \quad V = \frac{E}{Q}$$

V = Potential Difference (Voltage)
W = Work done
E = Energy
Q = Charge

Applications of Electric Potential

Electric potential is a fundamental concept in electromagnetism that measures the amount of electrical potential energy per unit charge at a point in an electric field. It has numerous practical applications in areas such as electrical power, electronics, medicine, and particle physics. Some practical applications of electric potential include:

- **Electrical power:** Electric potential is used to measure the voltage of an electrical circuit, which is a measure of the electrical potential difference between two points in the circuit. This is important in the generation and distribution of electrical power, where voltage needs to be controlled to ensure that electrical devices operate safely and efficiently.
- **Electronics:** In electronic circuits, the electric potential difference is used to control the flow of electric current, which is necessary for the operation of electronic devices such as transistors, diodes, and integrated circuits.
- **Medical applications:** Electric potential is used in medical applications such as electrocardiography (ECG), which measures the electrical activity of the heart, and electroencephalography (EEG), which measures the electrical activity of the brain.
- **Particle accelerators:** Electric potential is used to accelerate charged particles in particle accelerators, such as the Large Hadron Collider (LHC) at CERN, to very high speeds.

43. El-Sayed Rule

The El-Sayed rule, also known as the Kasha rule or Kasha's exciton model, is a principle that describes the fluorescence and phosphorescence of molecules. The rule was proposed by the Egyptian chemist Mostafa El-Sayed and the Polish-American physical chemist Michael Kasha in the 1950s.

According to the El-Sayed rule, the fluorescence and phosphorescence of a molecule are determined by the energy of the excited state of the molecule. Specifically, the rule states that fluorescence occurs when a molecule relaxes from a singlet excited state to the ground state, while phosphorescence occurs when a molecule relaxes from a triplet excited state to the ground state.

In more technical terms, the El-Sayed rule states that the probability of fluorescence is high when the excited state is a singlet state (S1), while the probability of phosphorescence is high when the excited state is a triplet state (T1). This is because singlet excited states have a higher probability of radiative decay (i.e. emitting light) than triplet excited states, which have a higher probability of non-radiative decay (i.e. releasing energy as heat).

The El-Sayed rule has important implications in a variety of fields to understand the electronic properties of molecules, and it has practical applications in the design of fluorescent and phosphorescent materials for use in organic light-emitting diodes (OLEDs), solar cells, and bio-imaging.

Mostafa El-Sayed

(Credit: https://commons.wikimedia.org/wiki/File:MAE_Profile_I_031516.jpg)

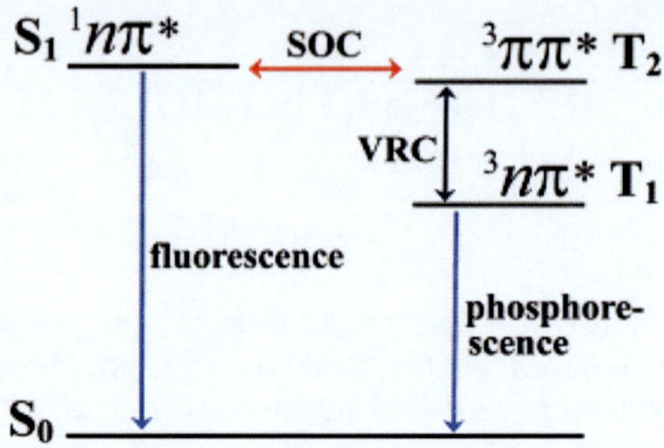

Applications of El-Sayed Rule

El-Sayed rule has a wide range of applications in photochemistry, materials science, and biological research. El-Sayed rule is a useful empirical relationship in photochemistry that relates the fluorescence lifetime of a molecule to its molecular structure. It provides a useful tool for probing the structure and properties of molecules, and has led to the development of new materials and technologies. Here are some applications of the El-Sayed rule:

- **Probing molecular structure:** The El-Sayed rule can be used to probe the molecular structure of a compound by measuring its fluorescence lifetime. By comparing the measured fluorescence lifetime to the predictions of the El-Sayed rule, researchers can infer information about the molecular structure of the compound.
- **Developing new materials:** The El-Sayed rule can be used in the development of new materials, such as fluorescent dyes and sensors. By designing molecules with specific structural features, researchers can control their fluorescence lifetimes and develop materials with unique optical properties.
- **Studying reaction dynamics:** The El-Sayed rule can be used to study reaction dynamics in photochemistry. By measuring the fluorescence lifetimes of reactants and products, researchers can study the mechanisms of photochemical reactions and gain insights into the factors that influence reaction rates.
- **Probing the properties of biological molecules:** The El-Sayed rule can also be used to study the properties of biological molecules, such as proteins and DNA. By labeling these molecules with fluorescent probes and measuring their fluorescence lifetimes, researchers can gain insights into their structure and dynamics.

44. EQUIVALENCE PRINCIPLE

The Equivalence Principle is a fundamental principle in physics that states that the effects of gravity are indistinguishable from the effects of acceleration. In other words, an observer in a gravitational field cannot tell whether they are being accelerated or are at rest in a gravitational field.

Albert Einstein

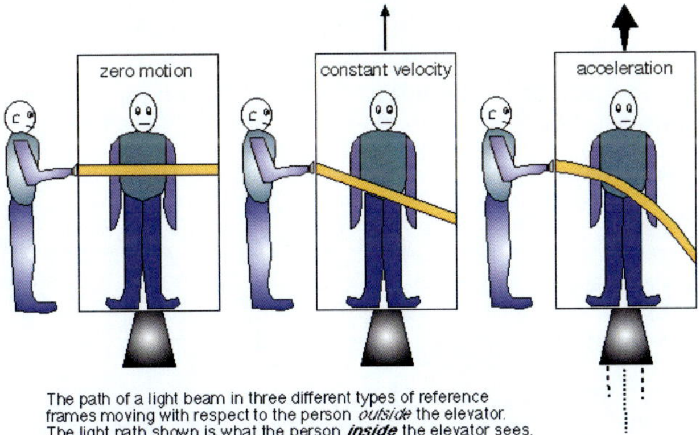

The path of a light beam in three different types of reference frames moving with respect to the person *outside* the elevator. The light path shown is what the person ***inside*** the elevator sees. Under large acceleration, the beam of light will curve downward. It should also do that in a region of strong gravity.

(Strobel, Nick. 2001. "Curved Spacetime". Accessed May 28, 2001)

Applications of Equivalence Principle

Equivalence Principle is a fundamental concept in physics that has many applications in various fields of science and engineering. It provides a powerful tool for understanding the behavior of gravity and has enabled the development of many important technologies, including interferometry and pulsar timing. The Equivalence Principle has many applications in physics, including:

- **General relativity:** The Equivalence Principle is a key concept in Albert Einstein's theory of General Relativity, which describes the behavior of gravity in terms of the curvature of space-time. According to General Relativity, gravity is not a force that pulls objects together, but rather a curvature of space-time that causes objects to move on curved paths.
- **Gravitational waves:** The Equivalence Principle is important in the study of gravitational waves, which are ripples in the fabric of space-time caused by the acceleration of massive objects. The Equivalence Principle ensures that the effects of gravitational waves can be detected using a variety of different techniques, including interferometry and pulsar timing.
- **Experimental physics:** The Equivalence Principle is important in experimental physics, particularly in tests of General Relativity and other theories of gravity. By carefully measuring the behavior of objects in a gravitational field, physicists can test the predictions of different theories of gravity and determine whether the Equivalence Principle holds true in different contexts.
- **Cosmology:** The Equivalence Principle is important in cosmology, the study of the structure and evolution of the universe. Cosmologists use the Equivalence Principle to study the effects of gravity on the large-scale structure of the universe, including the distribution of galaxies and the cosmic microwave background radiation.

45. EULER–LAGRANGE EQUATIONS

The Euler-Lagrange equations are a set of partial differential equations that describe the motion of a physical system based on its Lagrangian. They were developed by the Swiss mathematician Leonhard Euler and the Italian-French mathematician Joseph-Louis Lagrange in the 18th century.

The Lagrangian of a physical system is defined as the difference between the kinetic energy and the potential energy of the system. It is a function of the generalized coordinates of the system, parameters that describe the configuration of the system, and their time derivatives.

The Euler-Lagrange equations describe the evolution of the generalized coordinates of the system over time by minimizing the action of the system, which is the integral of the Lagrangian over time. The equations are given by:

$$\partial L/\partial q - d/dt(\partial L/\partial q') = 0$$

Here, q represents the generalized coordinates of the system, q' represents their time derivatives, L represents the Lagrangian of the system, and $\partial L/\partial q$ and $\partial L/\partial q'$ represent the partial derivatives of the Lagrangian with respect to the generalized coordinates and their time derivatives, respectively.

These equations can be used to derive the equations of motion of a wide range of physical systems, including particles, fields, and continuous media.

Leonhard Euler **Joseph-Louis Lagrange**

(Credit: Painting by Joseph Friedrich August Darbes; https://commons.wikimedia.org/wiki/File:Euler_1778.jpg)

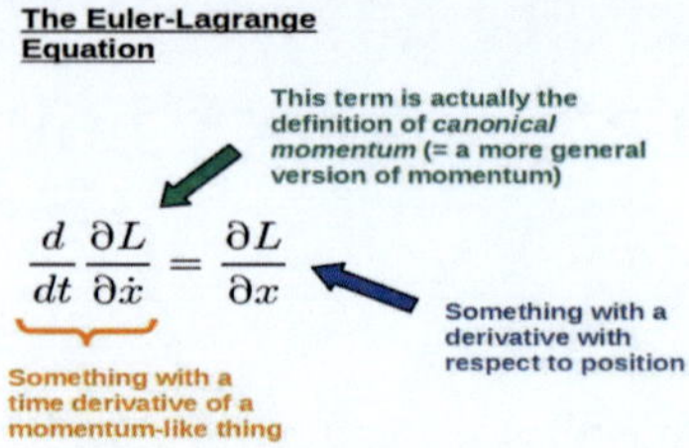

Applications of Euler–Lagrange Equations

The Euler-Lagrange equations are fundamental equations in classical mechanics that describe the motion of a system under the influence of potential and kinetic energies. They are used to find the path of least action, which is the path that minimizes the action integral.

- **Studying the motion of particles:** The Euler-Lagrange equations can be used to study the motion of particles in a wide range of physical systems. By understanding the forces acting on the particle and the resulting motion, researchers can predict the behavior of the system.
- **Analyzing the behavior of mechanical systems:** The Euler-Lagrange equations can be used to analyze the behavior of complex mechanical systems, such as robots and machines. By modeling the system as a set of interacting particles and applying the Euler-Lagrange equations, engineers can predict the motion of the system under different conditions.
- **Developing control systems:** The Euler-Lagrange equations are used in the development of control systems for various applications, such as robotics and aerospace. By understanding the forces acting on a system and the resulting motion, engineers can design control systems that manipulate the system's behavior to achieve a desired outcome.
- **Understanding the behavior of fluids:** The Euler-Lagrange equations can also be used to study the behavior of fluids, such as air and water. By modeling the fluid as a set of interacting particles and applying the Euler-Lagrange equations, researchers can predict the behavior of the fluid under different conditions.
- **Analyzing the behavior of quantum systems:** The Euler-Lagrange equations can be used to analyze the behavior of quantum systems, such as atoms and molecules. By understanding the forces acting on the system and the resulting motion, researchers can predict the behavior of the system under different conditions.

46. EULER'S EQUATIONS
(RIGID BODY DYNAMICS)

Euler's equations of rigid body dynamics are a set of equations that describe the rotational motion of a rigid body about a fixed point, also known as the center of mass. These equations were developed by the Swiss mathematician and physicist Leonhard Euler in the 18th century.

The three Euler's equations are:

$$T_x = I_x \dot{\omega}_x + (I_z - I_y)\,\omega_y\,\omega_z$$
$$T_y = I_y \dot{\omega}_y + (I_x - I_z)\,\omega_x\,\omega_z$$
$$T_z = I_z \dot{\omega}_z + (I_y - I_x)\,\omega_x\,\omega_y$$

Here, T_x, T_y, and T_z are the external torques acting about the x, y, and z axes, respectively; I_x, I_y, and I_z are the moments of inertia of the rigid body about the x, y, and z axes, respectively; ω_x, ω_y, and ω_z are the components of the angular velocity vector of the rigid body about the x, y, and z axes, respectively; and $\dot{\omega}_x$, $\dot{\omega}_y$, and $\dot{\omega}_z$ are the components of the angular acceleration vector of the rigid body about the x, y, and z axes, respectively.

These equations are derived from the principles of conservation of angular momentum and the rotational analog of Newton's second law, which states that the torque acting on an object is equal to its moment of inertia times its angular acceleration.

Euler's equations are used in a variety of fields, including aerospace engineering, robotics, and physics. They are used to model the motion of spacecraft, satellites, and other rigid bodies in space, as well as to design control systems for robotic manipulators and other mechanical systems.

Leonhard Euler

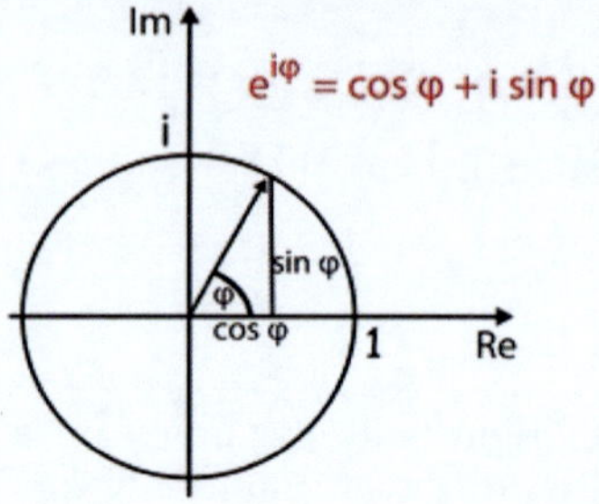

Applications of Euler's Equations

Euler's equations are a set of differential equations that describe the motion of a rigid body in three-dimensional space. They are fundamental principles in classical mechanics and have many important applications in physics and engineering. Here are some applications of Euler's equations:

- **Studying the motion of spacecraft:** Euler's equations can be used to study the motion of satellites and spacecraft in deep space. By modeling the spacecraft as a rigid body and applying Euler's equations, researchers can predict the orientation and motion of the spacecraft under the influence of external forces.
- **Designing aircraft:** Euler's equations are used in the design of aircraft, such as airplanes and helicopters. By understanding the motion of a rotating blade or wing, engineers can design aircraft that are stable, maneuverable, and efficient.
- **Studying fluid dynamics:** Euler's equations are also used in the study of fluid dynamics, which is the study of the motion of fluids, such as air and water. By applying Euler's equations to fluids, researchers can understand how they flow and interact with objects in their path.
- **Developing control systems:** Euler's equations are used in the development of control systems for various applications, such as robotics and aerospace. By understanding the motion of a rigid body and the forces acting on it, engineers can design control systems that manipulate the system's behavior to achieve a desired outcome.
- **Analyzing the behavior of spinning objects:** Euler's equations can be used to analyze the behavior of spinning objects, such as tops and gyroscopes. By understanding the motion of a spinning object and the forces acting on it, researchers can predict its behavior and develop new applications.

47. EULER'S LAWS OF MOTION

Euler's laws of motion are a set of three principles that describe the behavior of rigid bodies in motion. These laws were first proposed by the Swiss mathematician and physicist Leonhard Euler in the 18th century. In classical mechanics, Euler's laws of motion are equations of motion which extend Newton's laws of motion for point particle to rigid body motion.

The three laws are:

- **The first law:** Everybody persists in its state of being at rest or of moving uniformly straight forward, except insofar as it is compelled to change its state by forces impressed.

This law is also known as the law of inertia. It states that a body at rest will remain at rest, and a body in motion will continue to move at a constant velocity in a straight line, unless acted upon by an external force.

- **The second law:** The alteration of motion is ever proportional to the motive force impressed; and is made in the direction of the right line in which that force is impressed.

This law is also known as the law of acceleration. It states that the acceleration of a body is directly proportional to the force applied to it, and inversely proportional to its mass. This can be expressed mathematically as $F = ma$, where F is the force, m is the mass, and a is the acceleration.

- **The third law:** To every action there is always opposed an equal reaction; or, the mutual actions of two bodies upon each other are always equal, and directed to contrary parts.

This law is also known as the law of action and reaction. It states that for every action (force) there is an equal and opposite reaction (force). In other words, if one object exerts a force on another object, the second object will exert an equal and opposite force on the first object.

Leonhard Euler

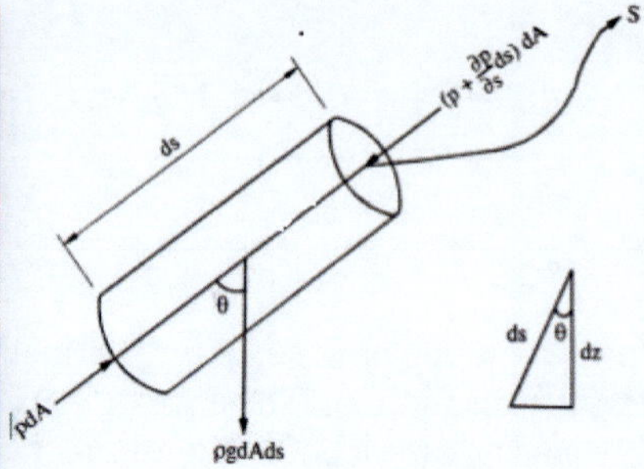

Applications of Euler's Laws of Motion

Euler's laws of motion are fundamental principles in classical mechanics that describe the motion of objects under the influence of forces. Here are some applications of Euler's laws of motion:

- **Designing vehicles:** Euler's laws of motion can be used in the design of vehicles, such as airplanes, cars, and spacecraft. By understanding the forces acting on the vehicle and the resulting motion, engineers can design vehicles that are safe, efficient, and maneuverable.
- **Understanding fluid dynamics:** Euler's laws of motion are also used in the study of fluid dynamics, which is the study of the motion of fluids, such as air and water. By applying the laws of motion to fluids, researchers can understand how they flow and interact with objects in their path.
- **Studying the behavior of celestial bodies:** Euler's laws of motion are used to study the behavior of celestial bodies, such as planets and asteroids. By applying the laws of motion to these bodies, researchers can understand their orbits, gravitational interactions, and other aspects of their motion.
- **Developing control systems:** Euler's laws of motion are also used in the development of control systems for various applications, such as robotics and aerospace. By understanding the forces acting on a system and the resulting motion, engineers can design control systems that manipulate the system's behavior to achieve a desired outcome.
- **Modeling complex systems:** Euler's laws of motion can be used to model complex systems, such as the behavior of particles in a gas or the motion of a flock of birds. By simulating the motion of individual particles under the influence of forces, researchers can understand the behavior of the system as a whole.

48. FARADAY'S LAW

Faraday's law of electromagnetic induction is a fundamental law in electromagnetism, discovered by Michael Faraday, a British physicist and chemist, in the early 19th century. The law states that a changing magnetic field in a closed loop of wire induces an electromotive force (EMF) in the loop, which in turn generates an electric current.

The law is expressed mathematically as:

$$\mathbf{EMF = -\,d\Phi\,/\,dt}$$

where EMF is the electromotive force induced in the circuit (measured in volts).

$d\Phi/dt$ represents the rate of change of magnetic flux through the circuit with respect to time (measured in webers per second, or tesla per second).

The negative sign indicates that the induced EMF creates a current that opposes the change in magnetic flux.

Faraday's law of electromagnetic induction states that a change in the magnetic field within a closed loop of wire induces an electromotive force (EMF) or voltage across the wire. In other words, when the magnetic field through a conductor changes, it induces an electric current in the conductor.

This law is essential in understanding how generators and transformers work, as it provides the basis for generating electricity through electromagnetic induction.

Michael Faraday

Faraday's Law Equation

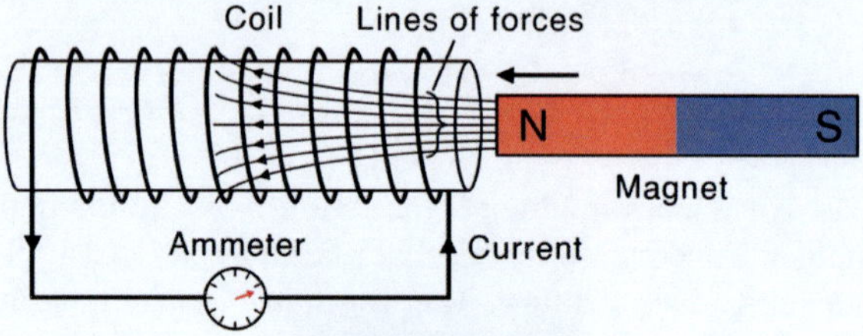

$$\varepsilon = -N \frac{d\phi}{dt}$$

ε : Electromotive force (EMF)

N : Number of turns of the coil

$\frac{d\phi}{dt}$: Instaneous change of magnetic flux with time

Applications of Faraday's Law

Faraday's law is a crucial principle in electromagnetism, with numerous practical applications in various fields of physics and engineering. Faraday's law has many practical applications, including:

◆ **Electrical power generation:** Electric generators use Faraday's law to convert mechanical energy into electrical energy by rotating a coil of wire in a magnetic field.

◆ **Transformers:** Transformers use Faraday's law to transfer electrical power from one circuit to another by inducing an EMF in a secondary coil through a changing magnetic field created by an alternating current in the primary coil.

◆ **Induction heating:** Faraday's law is used in induction heating, which uses a changing magnetic field to heat conductive materials.

◆ **Magnetic braking:** Faraday's law is used in magnetic braking, which uses the magnetic force generated by an induced current to slow down or stop a moving object.

◆ **Electromagnetic compatibility testing:** Faraday's law is used in electromagnetic compatibility testing, which evaluates the ability of electronic devices to operate without interference from other devices.

49. FARADAY'S LAWS OF ELECTROLYSIS

Faraday's laws of electrolysis are a set of fundamental laws governing the behavior of electrolytic cells. These laws were discovered by Michael Faraday in the 1830s and are named after him.

The two laws are as follows:

➢ **Faraday's first law:** The amount of substance produced at an electrode during electrolysis is directly proportional to the quantity of electricity passed through the electrolyte.

This law states that the amount of a substance (m) produced at an electrode during electrolysis is proportional to the amount of electrical charge (or quantity of electricity) that passes through the electrolyte. This law is expressed mathematically as:

$$m = Z\,I\,t$$

where,

Z = electrochemical equivalent of the substance

I = electric current

t = time

The electrochemical equivalent of a substance is the amount of substance produced per unit of electrical charge passed through the electrolyte.

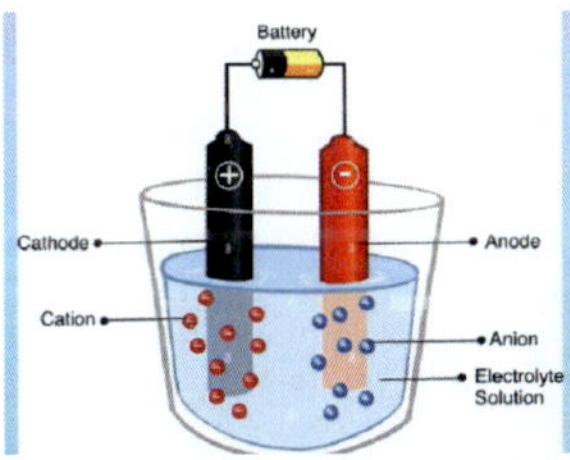

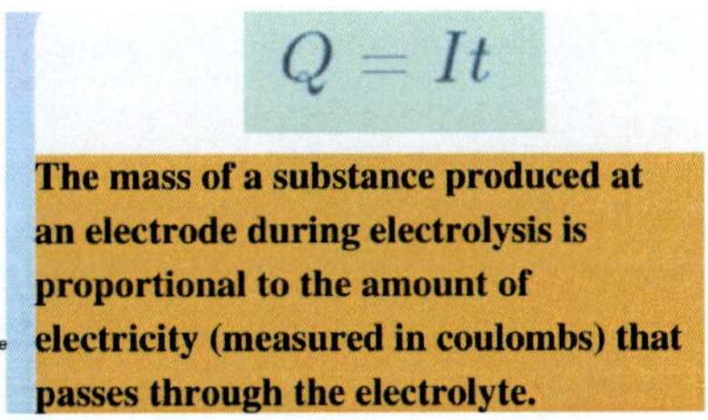

➢ **Faraday's second law:** The amounts of different substances produced by the same quantity of electricity passing through an electrolyte are proportional to their chemical equivalent weights.

This law states that the ratio of the masses of two different substances produced by the same amount of electricity passing through the electrolyte is equal to the ratio of their chemical equivalent weights. This law is expressed mathematically as:

$$m_1 / m_2 = E_1 / E_2$$

where, m_1 and m_2 are the masses of the two substances produced

E_1 and E_2 are the chemical equivalent weights of the two substances

The chemical equivalent weight of a substance is the mass of the substance that is chemically equivalent to one mole of electrons.

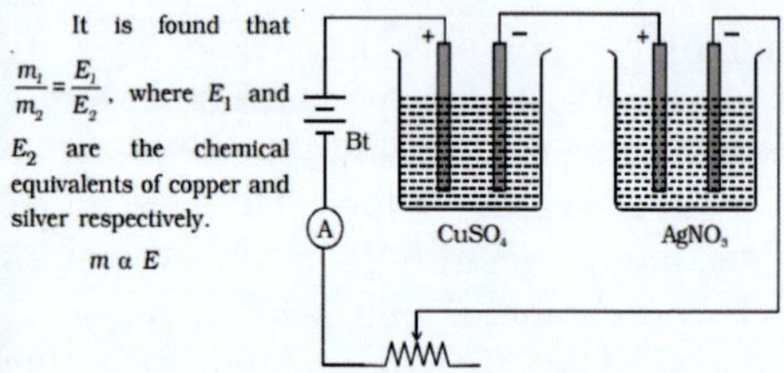

Michael Faraday

(Credit: https://wellcomecollection.org/works/tsejjkhu)

Applications of Faraday's Laws of Electrolysis

Faraday's laws of electrolysis have several important applications in chemistry and industry. Some of the main applications are:

- **Electroplating:** Faraday's laws are used to control the amount of metal deposited during electroplating. Electroplating is the process of coating a metal object with a thin layer of another metal using electrolysis. By controlling the current and time, the thickness and uniformity of the coating can be controlled.
- **Purification of metals:** Faraday's laws are also used to purify metals. Electrolytic refining is a process in which impure metals are purified using electrolysis.
- **Production of chemicals:** Electrolysis is used to produce many important chemicals such as chlorine, sodium hydroxide, and aluminum. Faraday's laws are used to determine the amount of electricity required to produce a given amount of the desired product.
- **Quantitative analysis:** Faraday's laws can be used to determine the amount of a substance present in a sample. This is done by measuring the amount of electricity required to produce a known amount of the substance during electrolysis.

50. FAXÉN'S LAW

Faxén's law, named after the Swedish physicist Hilding Faxén, describes the force experienced by a small particle suspended in a fluid due to its motion through that fluid. The law provides an approximation for calculating the force on a small particle moving through a viscous fluid.

The general form of Faxén's law for a spherical particle moving through a viscous fluid is given by:

$$\mathbf{F = 6\pi\mu a\ (v - V_p)}$$

where,

F is the force on the particle,

μ is the dynamic viscosity of the fluid,

a is the particle radius,

v is the velocity of the particle, and

V_p is the velocity of the fluid at the particle's location.

Faxén's law considers two main effects: the drag force and the added mass force. The drag force, 6πμav, represents the resistance experienced by the particle due to its motion through the fluid. The added mass force, $6\pi\mu aV_p$, accounts for the inertia of the surrounding fluid displaced by the particle's motion.

By using Faxén's law, you can calculate the force on the particle, given the particle's velocity and the fluid's velocity at the particle's location. This force can then be used to analyze the motion and behavior of the particle in the fluid.

It's important to note that Faxén's law is an approximation and assumes certain conditions, such as laminar flow and low Reynolds number. The Reynolds number (Re) is a dimensionless quantity that compares the inertial forces to viscous forces in the fluid. For more complex situations or when these assumptions do not hold (e.g. if Re is high), alternative models or approaches, such as Stokes' law or the drag coefficient approach, may be more appropriate.

Faxén's law is important in the study of sedimentation, which is the process by which particles settle out of a fluid under the influence of gravity. The law helps to predict the settling behavior of particles in a fluid and is used in the design of industrial processes such as sedimentation tanks and centrifuges.

Applications of Faxén's Law

Faxén's law is a fundamental law in fluid mechanics that describes the force acting on a small object moving through a fluid. The law was first proposed by the Swedish mathematician and physicist Hilding Faxén in the early 20th century. Faxén's law has many important applications in fluid mechanics, biophysics, and engineering. It has led to many important technological advancements. Here are some applications of Faxén's law:

- **Studying the behavior of microorganisms:** Faxén's law is used in the study of microorganisms, such as bacteria and algae, that move through fluids. By understanding the forces acting on the microorganisms and the resulting motion, researchers can predict the behavior of the microorganisms and develop new treatments for diseases caused by them.
- **Designing microfluidic devices:** Faxén's law is used in the design of microfluidic devices, which are devices that manipulate small volumes of fluids. By understanding the forces acting on small objects in the fluid, engineers can design microfluidic devices that are more efficient and effective.
- **Studying the behavior of sediment particles:** Faxén's law is used in the study of sediment particles, such as sand and silt, that move through fluids. By understanding the forces acting on the sediment particles and the resulting motion, researchers can predict the behavior of sediment particles in natural systems, such as rivers and oceans.
- **Understanding the behavior of colloidal particles**: Faxén's law is used in the study of colloidal particles, which are small particles that are suspended in a fluid. By understanding the forces acting on the colloidal particles and the resulting motion, researchers can predict the behavior of the colloidal particles in various systems, such as biological systems and industrial processes.

51. FERMAT'S PRINCIPLE

Fermat's principle, also known as the principle of least time, is a fundamental principle in optics that describes how light waves propagate through space. It states that light takes the path between two points that requires the least time, compared to all other possible paths.

Mathematically, Fermat's principle can be expressed as follows: the path taken by a light ray between two points is such that the optical path length is stationary. The optical path length is defined as the product of the distance traveled by the light ray and the refractive index of the medium through which it is traveling.

Pierre de Fermat

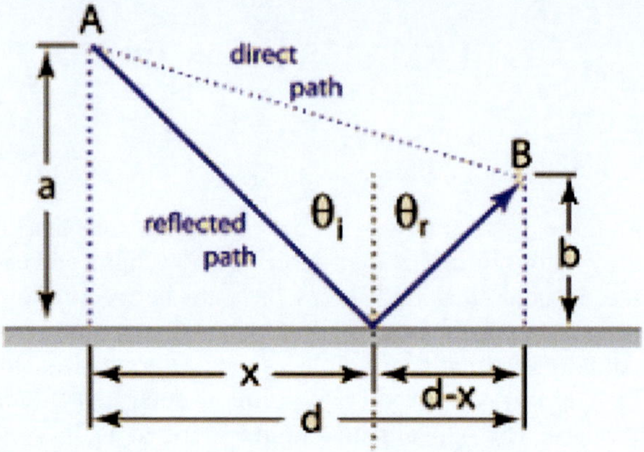

Applications of Fermat's Principle

Fermat's principle has many applications in optics and engineering, including:

♦ **Refraction:** Fermat's principle is used to describe the path taken by light rays as they pass through different media with varying refractive indices. It explains why light bends as it passes through a prism, and why lenses are able to focus light.

♦ **Reflection:** Fermat's principle is used to describe the path taken by light rays as they are reflected off surfaces, such as mirrors. It explains why the angle of incidence equals the angle of reflection.

♦ **Geometrical optics:** Fermat's principle is used in geometrical optics to predict the path taken by light rays in various optical systems, such as lenses and mirrors.

♦ **Optimal path finding:** Fermat's principle has applications beyond optics and is used in other fields of science and engineering to find the optimal path between two points. For example, it can be used to optimize the path of a rocket or a particle in a magnetic field.

Fermat's principle is a fundamental principle in optics that has numerous applications in various fields of physics and engineering. It provides a powerful tool for understanding the behavior of light waves and has enabled the development of many important technologies, such as lenses, mirrors, and optical fibers.

52. FERMI PARADOX

The Fermi paradox is the apparent contradiction between the high probability of the existence of extraterrestrial civilizations and the lack of evidence for or contact with such civilizations. The paradox is named after Enrico Fermi, who first posed the question, "Where is everybody?" in 1950. There are a number of proposed solutions to the Fermi paradox, including:

- **The Rare Earth hypothesis:** This hypothesis suggests that Earth-like planets are rare and that the conditions necessary for the development of intelligent life are unlikely to occur elsewhere in the universe.
- **The Great Filter hypothesis:** This hypothesis suggests that there is a "filter" that prevents the development of intelligent life beyond a certain point. This filter could be any number of factors, such as natural disasters, biological or technological factors, or environmental factors.
- **The Zoo hypothesis:** This hypothesis suggests that intelligent extraterrestrial civilizations are aware of Earth's existence, but have chosen not to make contact with us for various reasons, such as a desire to observe without interference or to prevent cultural contamination.
- **The Self-Destructive hypothesis:** This hypothesis suggests that intelligent civilizations inevitably destroy themselves through war, environmental damage, or other factors before they are able to establish contact with other civilizations.

Enrico Fermi

Applications of Fermi Paradox

The Fermi paradox does not have direct practical applications, as it is a theoretical concept in astrobiology and does not have immediate technological or scientific applications. However, the Fermi paradox has inspired a number of thought experiments and discussions that have contributed to our understanding of the universe and our place in it. Some of the indirect applications of the Fermi paradox include:

- **The search for extraterrestrial intelligence (SETI):** The Fermi paradox has motivated the search for extraterrestrial intelligence, as it raises the question of why we have not yet detected any signs of other intelligent civilizations. The SETI program searches for evidence of intelligent life in the universe through the detection of signals from other planets or civilizations.
- **Astrobiology:** The Fermi paradox is a key concept in astrobiology, which is the study of the origin, evolution, distribution, and future of life in the universe. The paradox raises important questions about the conditions necessary for the development of intelligent life and the likelihood of its existence elsewhere in the universe.
- **Philosophy and ethics:** The Fermi paradox has inspired philosophical and ethical discussions about the nature of life, the future of humanity, and our place in the universe. It has raised questions about the value of life, the potential for technological progress, and the ethical responsibilities of advanced civilizations.

Fermi paradox has contributed to our understanding of the universe and our place in it, and has motivated important scientific and philosophical discussions. While it does not have direct practical applications, it has inspired research and exploration that could have important implications for our future as a species.

53. FICK'S LAW OF DIFFUSION

Fick's law of diffusion is a scientific principle that describes the rate at which particles diffuse through a medium. Specifically, it states that the rate of diffusion is proportional to the concentration gradient of the particles and the area through which the particles are diffusing, and inversely proportional to the thickness of the medium through which the particles are diffusing. In mathematical terms, Fick's law can be expressed as:

$$J = - D (dC/dx)$$

where J is the flux of particles (i.e., the amount of particles that cross a unit area per unit time), D is the diffusion coefficient (which depends on the properties of the medium and the particles), C is the concentration of particles, x is the distance, and dC/dx is the concentration gradient (i.e., the change in concentration over a unit distance). The negative sign indicates that the flux is in the direction of decreasing concentration (i.e., from high to low concentration).

Adolf Eugen Fick

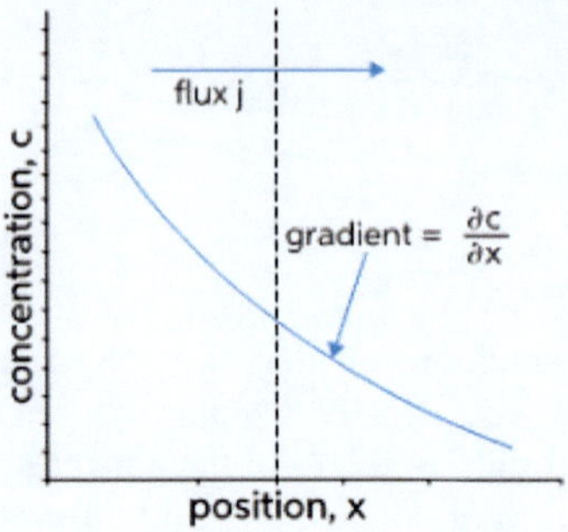

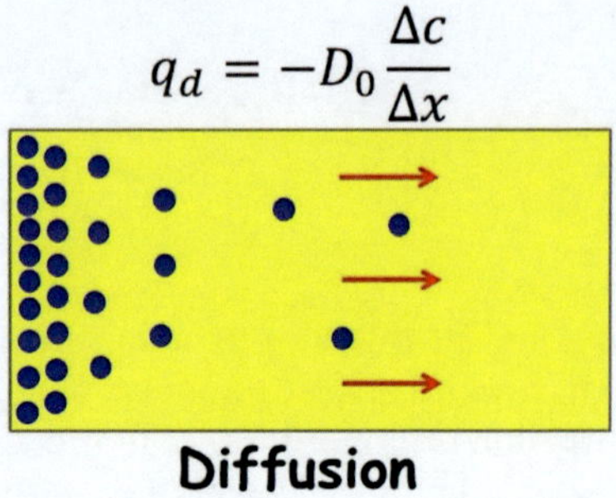

Applications of Fick's Law of Diffusion

Fick's law is used in a wide variety of fields, including physics, chemistry, biology, and engineering, to model diffusion processes.

- **Biology:** Fick's law of diffusion is used to describe the movement of molecules across cell membranes. It provides a framework for understanding how nutrients, gases, and other molecules are transported into and out of cells.
- **Chemistry:** Fick's law of diffusion is used to study the diffusion of molecules in various chemical reactions. This information is useful for predicting the rate of reaction and for optimizing reaction conditions.
- **Environmental science:** Fick's law of diffusion is used to study the movement of pollutants through soil and groundwater. It provides a framework for understanding how pollutants move through different types of soil and how they can be remediated.
- **Materials science:** Fick's law of diffusion is used to study the diffusion of atoms and molecules in materials. This information is useful for predicting the rate of diffusion and for designing materials with specific properties.
- **Pharmacy:** Fick's law of diffusion is used in the study of drug delivery systems, such as transdermal patches and inhalers. It provides a framework for understanding how drugs are released from these systems and how they are absorbed by the body.

Fick's law of diffusion is a fundamental concept in many fields and is used to study the movement of molecules in different types of systems. It provides important information about the rate and direction of diffusion, which is essential for understanding the behavior of many different physical and chemical systems.

54. First Law of Thermodynamics

The first law of thermodynamics, also known as the law of conservation of energy, states that energy cannot be created or destroyed, only transformed from one form to another. In other words, the total amount of energy in a closed system remains constant, but it may be converted from one form to another, such as from heat to work or from potential energy to kinetic energy. Mathematically, the first law of thermodynamics can be expressed as:

$$\Delta U = Q - W$$

where ΔU is the change in internal energy of the system, Q is the heat added to the system, and W is the work done by the system. This equation shows that the change in internal energy of the system is equal to the heat added to the system minus the work done by the system.

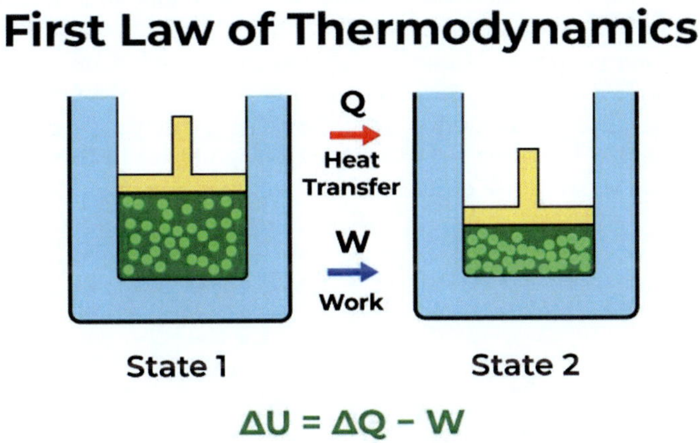

Applications of First Law of Thermodynamics

The first law of thermodynamics is a fundamental principle of energy conservation that has many practical applications in various fields. Some of the important applications of the first law of thermodynamics are:

- **Heat engines:** The first law of thermodynamics is used to study the efficiency and performance of heat engines, which convert thermal energy into mechanical work. The law helps to determine the amount of heat input required to produce a given amount of work, and to calculate the efficiency of the engine.
- **Refrigeration and air conditioning:** The first law of thermodynamics is used to study the cooling and refrigeration processes, which involve the transfer of heat from a low-temperature environment to a high-temperature environment. The law helps to determine the amount of work input required to remove heat from a system and to maintain a desired temperature.
- **Chemical reactions:** The first law of thermodynamics is used to study chemical reactions and to determine the amount of energy released or absorbed during the reaction. The law helps to calculate the enthalpy change (ΔH) of a reaction, which is the difference between the energy of the products and the energy of the reactants.
- **Power generation:** The first law of thermodynamics is used to study power generation processes, such as steam turbines and gas turbines, which convert heat energy into electricity. The law helps to determine the amount of heat input required to produce a given amount of power, and to calculate the efficiency of the power plant.
- **Energy conservation:** The first law of thermodynamics is used to study and promote energy conservation in various systems, such as buildings, vehicles, and industrial processes. The law helps to identify areas where energy is being wasted and to develop strategies to reduce energy consumption and improve efficiency.

55. FOURIER'S LAW

Fourier's Law, named after French mathematician and physicist Joseph Fourier, describes the heat transfer in a system through conduction. The law states that the rate of heat transfer through a material is directly proportional to the temperature gradient, and is given by the equation:

$$q = -kA\,(dT/dx)$$

where,

q = the heat flow rate, measured in watts (W)

k = the thermal conductivity of the material, measured in watts per meter per kelvin (W/mK)

A = the cross-sectional area of the material through which heat is flowing, measured in square meters (m^2)

dT/dx = the temperature gradient, measured in kelvins per meter (K/m)

The negative sign in the equation indicates that heat flows from high temperature to low temperature. In other words, heat will flow in the direction of decreasing temperature.

Joseph Fourier

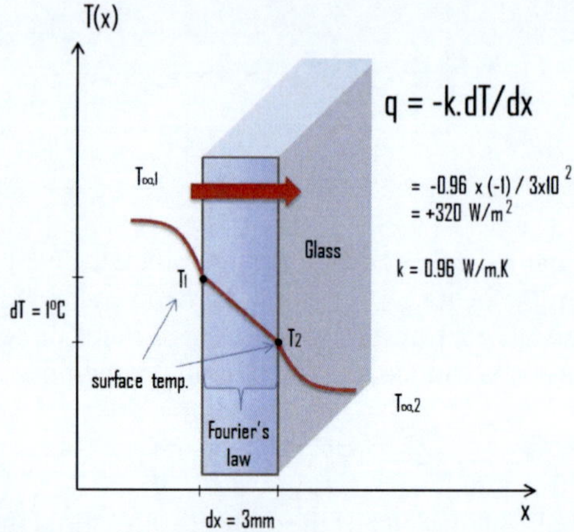

Application of Fourier's Law

Fourier's Law has a wide range of applications. The law is widely used in engineering and physics to describe heat transfer in various applications, such as in buildings, electronics, and power plants.

- **Heat transfer in buildings:** Fourier's Law is used to determine the amount of heat transferred through walls, roofs, and floors of buildings, which helps in designing efficient heating and cooling systems.
- **Thermal management of electronic devices:** The law is used to design cooling systems for electronic devices like computer processors, which generate a lot of heat. The heat transfer through the cooling system is determined using Fourier's Law.
- **Power generation:** Fourier's Law is used to design heat exchangers for power plants, which transfer heat from one fluid to another. The law is also used to determine the thermal performance of nuclear reactors.
- **Materials science:** Fourier's Law is used to study the thermal conductivity of various materials. This helps in selecting materials for applications like insulators and conductors.
- **Food processing:** Fourier's Law is used to determine the rate of heat transfer in food processing applications such as baking, frying, and pasteurization.

Fourier's Law plays a vital role in understanding and designing systems that involve heat transfer, making it a fundamental concept in engineering and physics.

56. GAUSS' LAW

Gauss' law is a fundamental law of electromagnetism that relates the distribution of electric charges to the electric field they produce. The law states that the electric flux through any closed surface is proportional to the total electric charge enclosed within the surface.

Mathematically, Gauss' law can be written as:

$$\oint S\ \mathbf{E \cdot dA} = \mathbf{Q} / \varepsilon_0$$

where $\oint$S is the surface integral over a closed surface S, E is the electric field, dA is the differential area element of the surface, Q is the total electric charge enclosed within the surface, and ε_0 is the electric constant (also known as the permittivity of free space).

Carl Friedrich Gauss

Gauss's law Formula

$$d\Phi_E = \vec{\mathbf{E}} \bullet d\vec{\mathbf{A}} = E\, dA \cos\theta ,$$

$$d\Phi_E = \frac{q}{4\pi\varepsilon_0} \frac{dA \cos\theta}{r^2}$$

$$d\Phi_E = \frac{q}{4\pi\varepsilon_0} d\omega$$

$$\Phi_E = \oint \vec{\mathbf{E}} \bullet d\vec{\mathbf{A}} = \frac{q}{4\pi\varepsilon_0} \oint d\omega$$

$$\boxed{\Phi_E = \frac{q}{\varepsilon_0} .}$$

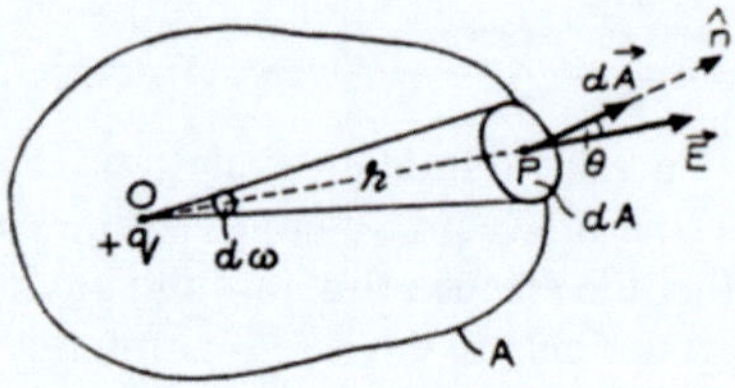

Applications of Gauss' Law

Some practical applications of Gauss' law include:

- **Electrostatics:** Gauss' law is used to calculate the electric field produced by a given charge distribution, and it is an important tool for solving problems in electrostatics.
- **Electrical engineering:** Gauss' law is used in the design and analysis of electrical circuits, such as capacitors and antennas.
- **Physics research:** Gauss' law is used in many areas of physics research, including the study of plasma physics, quantum mechanics, and astrophysics.
- **Materials science:** Gauss' law is used to understand the behavior of materials in the presence of an electric field, and it is important in the design of materials for applications such as capacitors and sensors.

In summary, Gauss' law is a fundamental law of electromagnetism that relates the distribution of electric charges to the electric field they produce. It has numerous practical applications in areas such as electrostatics, electrical engineering, physics research, and materials science.

57. GAY-LUSSAC'S LAW

Gay-Lussac's law, also known as the law of combining volumes, is an experimental gas law that describes the relationship between the pressure and temperature of a gas at constant volume.

The law states that, at constant volume, the pressure of a given mass of an ideal gas is directly proportional to its absolute temperature. Mathematically, Gay-Lussac's law can be expressed as:

$$P = k \times T$$

where P is the pressure of the gas, T is its absolute temperature (measured in kelvin), and k is a constant of proportionality.

This law was first discovered by the French chemist Joseph Louis Gay-Lussac in the early 19th century, based on his experiments with gases at different temperatures and pressures. Gay-Lussac's law is related to the kinetic theory of gases, which states that the pressure of a gas is related to the average kinetic energy of its particles.

Gay-Lussac's law is an important concept in the study of thermodynamics and is widely used in various fields of science and engineering, such as in the design of engines and refrigeration systems. It also helps to explain many everyday phenomena, such as why the pressure in a container increases when it is heated or why a tire might burst on a hot day.

Joseph Louis Gay-Lussac

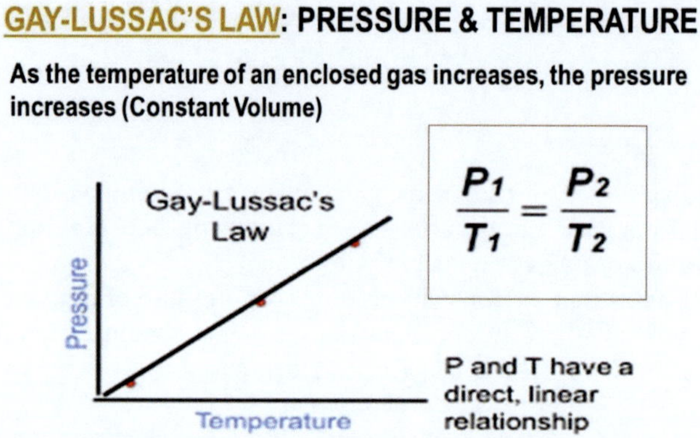

Applications of Gay-Lussac's Law

Gay-Lussac's law, also known as the law of combining volumes, states that when gases react, they do so in volumes that are in a ratio of small whole numbers if all the volumes are measured at the same temperature and pressure. This law has several applications, including:

- **Stoichiometry:** Gay-Lussac's law helps in determining the ratio of volumes of reactants and products in chemical reactions, which is useful in stoichiometric calculations.
- **Gas laws:** Gay-Lussac's law is one of the fundamental gas laws, along with Boyle's law and Charles's law, and helps in understanding the behavior of gases under different conditions.
- **Gas density:** The volume of a gas is directly proportional to the number of gas molecules present in the given volume. Therefore, Gay-Lussac's law is useful in determining the density of gases.
- **Avogadro's hypothesis:** Avogadro's hypothesis states that equal volumes of gases at the same temperature and pressure contain the same number of molecules. Gay-Lussac's law is used to demonstrate this hypothesis, as it shows that the ratio of the volumes of reactants and products is always in small whole numbers.
- **Gas mixtures:** Gay-Lussac's law is used to determine the composition of gas mixtures. The ratio of the volumes of the gases in a mixture can be used to determine the ratio of the number of molecules of each gas present.

58. GEM Equations

The GEM equations refer to the Gauss, Ampère, and Faraday laws of electromagnetism, which are a set of four partial differential equations that describe the behavior of electric and magnetic fields in space and time.

The equations are:

- **Gauss's law:** This law relates the electric field to the charge density. It states that the flux of the electric field through a closed surface is proportional to the charge enclosed within the surface.
- **Ampère's law:** This law relates the magnetic field to the current density. It states that the magnetic field around a closed loop is proportional to the current flowing through the loop.
- **Faraday's law:** This law describes the relationship between a changing magnetic field and an induced electric field. It states that an electric field is induced in any closed loop when the magnetic flux through the loop changes.
- **Gauss's law for magnetism:** This law relates the magnetic field to the magnetic charge density. It states that the flux of the magnetic field through any closed surface is always zero.

These four equations are known collectively as the Maxwell's equations, named after the physicist James Clerk Maxwell, who first formulated them in the 19th century. The equations are fundamental to the study of electromagnetism, and have wide-ranging applications in fields such as optics, communications, and electronics.

James Clerk Maxwell

$$\nabla \cdot \mathbf{D} = \rho \qquad (1) \qquad \text{Gauss' Law}$$

$$\nabla \cdot \mathbf{B} = 0 \qquad (2) \qquad \text{Gauss' Law for magnetism}$$

$$\nabla \times \mathbf{E} = -\frac{\partial \mathbf{B}}{\partial t} \qquad (3) \qquad \text{Faraday's Law}$$

$$\nabla \times \mathbf{H} = \frac{\partial \mathbf{D}}{\partial t} + \mathbf{J} \qquad (4) \qquad \text{Ampère-Maxwell Law}$$

Applications of GEM Equations

The GEM equations (Gravity, Electromagnetism, and Yang-Mills theory) are a set of equations that describe the behavior of gravitation and electromagnetism, as well as non-Abelian gauge fields in high-energy physics.

- One application of the GEM equations is in the study of cosmology and the early universe. In the early universe, the GEM equations can be used to study the behavior of the electromagnetic and gravitational fields during the period of inflation, which is thought to have occurred in the first fractions of a second after the Big Bang. This can provide insight into the origin of the large-scale structure of the universe.
- Another application of the GEM equations is in the study of black holes. The equations can be used to describe the behavior of the electromagnetic and gravitational fields in the vicinity of a black hole, which can help in understanding the properties and behavior of these objects.
- The GEM equations are also used in the study of high-energy physics, where they are used to describe the behavior of particles in strong electromagnetic and gravitational fields. This can help in understanding the behavior of particles in particle accelerators and in extreme astrophysical environments, such as neutron stars and black holes.

59. Geodesic Equation

The geodesic equation is a fundamental equation of general relativity that describes the motion of objects moving in a curved space-time. In particular, it describes the path followed by objects moving under the influence of gravity, and is the basis for the study of gravitational lensing and the motion of celestial objects such as planets, stars, and galaxies.

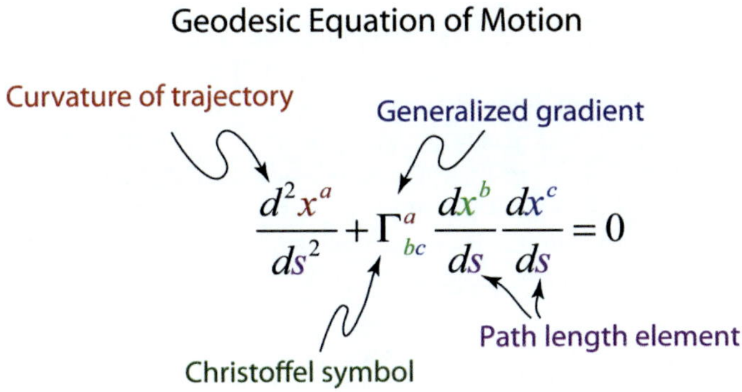

$$\frac{d^2x^a}{ds^2} + \Gamma^a_{bc}\frac{dx^b}{ds}\frac{dx^c}{ds} = 0$$

(Nolte, David. 2019. "How to Teach General Relativity to Undergraduate Physics Majors". Accessed July 15, 2019)

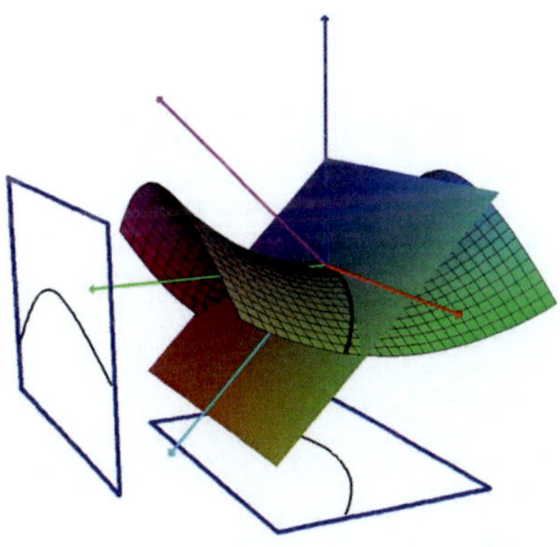

The geodesic equation is given by Christoffel symbols, which describe the curvature of space-time.

The geodesic equation is a second-order differential equation, and its solutions are the curves in space-time that are followed by objects moving freely under the influence of gravity. These curves are known as geodesics, and they correspond to the straightest possible paths in a curved space-time.

Applications of Geodesic Equation

The geodesic equation is a fundamental concept in general relativity that describes the motion of a test particle moving under the influence of gravity. It is the equation of motion that governs the trajectory of a free-falling particle in a curved space-time.

The geodesic equation has many applications, some of which include:

- **Orbit calculation:** The geodesic equation can be used to calculate the orbit of planets and other celestial bodies around a star or black hole. It is an important tool used in astrophysics and space exploration.
- **Gravitational lensing:** The geodesic equation helps explain the phenomenon of gravitational lensing, where the path of light is bent by the curvature of space-time. It is used to calculate the distortion of light caused by massive objects such as galaxies, which can be observed through telescopes.
- **Black holes:** The geodesic equation is essential in understanding the behavior of objects falling into a black hole. It can be used to calculate the path of a test particle as it approaches the event horizon and the singularity.
- **Global positioning systems (GPS):** The geodesic equation is used in the accurate positioning of GPS satellites and the calculation of the position of GPS receivers on the Earth's surface. It helps to correct for relativistic effects, such as time dilation, that affect the accuracy of GPS measurements.
- **Spacecraft navigation:** The geodesic equation is used in spacecraft navigation to determine the trajectory of a spacecraft in a gravitational field. It is also used to calculate the gravitational forces acting on a spacecraft and to determine the best trajectory for a mission.

60. GIBBS–HELMHOLTZ EQUATION

The Gibbs-Helmholtz equation, named after the American physicist Josiah Willard Gibbs and the German physicist Hermann von Helmholtz, is an important thermodynamic equation that relates the change in the Gibbs free energy of a system to the change in its temperature at constant pressure.

The Gibbs free energy, denoted by G, is a measure of the amount of energy available to do work in a thermodynamic system. It is defined as the difference between the enthalpy of the system, denoted by H, and the product of the temperature and the entropy of the system, denoted by TS.

The Gibbs-Helmholtz equation states that the change in the Gibbs free energy of a system, denoted by ΔG, with respect to temperature at constant pressure, denoted by ΔT, is related to the enthalpy, denoted by H, and the entropy, denoted by S, of the system by the following equation:

$$\Delta G/\Delta T = -\Delta H/(\Delta T)^\wedge 2 + \Delta S/\Delta T$$

Here, the derivatives are taken at constant pressure.

The Gibbs-Helmholtz equation is particularly useful for studying the thermodynamic stability of a system. It can be used to calculate the temperature at which a chemical reaction becomes spontaneous, as well as to predict the effect of temperature on the equilibrium constant of a chemical reaction. The equation is useful in the study of phase transitions in materials, such as the melting and boiling points of substances.

Josiah Willard Gibbs

Hermann von Helmholt

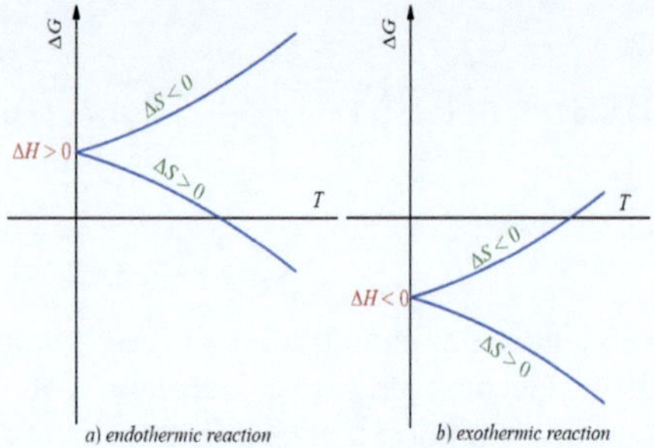

Applications of Gibbs–Helmholtz Equation

The Gibbs-Helmholtz equation is a fundamental equation in thermodynamics that relates the change in Gibbs free energy of a system to the change in temperature and pressure. Gibbs-Helmholtz equation has many important applications in thermodynamics, chemistry, and biology. Here are some applications of the Gibbs-Helmholtz equation:

- **Determining thermodynamic properties:** The Gibbs-Helmholtz equation can be used to determine the thermodynamic properties of a system, such as the enthalpy and entropy, at different temperatures and pressures. By understanding how these properties change with temperature and pressure, researchers can predict the behavior of the system under different conditions.
- **Designing chemical reactions:** The Gibbs-Helmholtz equation is used in the design of chemical reactions, such as the production of pharmaceuticals and industrial chemicals. By understanding the change in Gibbs free energy of the reaction with temperature and pressure, chemists can design reactions that are more efficient and effective.
- **Studying protein folding:** The Gibbs-Helmholtz equation is used in the study of protein folding, which is the process by which a protein assumes its final three-dimensional structure. By understanding the change in Gibbs free energy of the protein with temperature and pressure, researchers can predict the stability of the protein and the conditions under which it will fold correctly.
- **Analyzing phase transitions:** The Gibbs-Helmholtz equation is used in the analysis of phase transitions, such as the transition from liquid to gas or solid to liquid. By understanding the change in Gibbs free energy with temperature and pressure, researchers can predict the conditions under which a phase transition will occur.

61. GRAHAM'S LAW

Graham's law, named after the Scottish chemist Thomas Graham, relates the rates of effusion or diffusion of two gases under the same conditions of temperature and pressure.

The rate of effusion or diffusion is the rate at which a gas escapes through a small hole or mixes with another gas, respectively. According to Graham's law, the rate of effusion or diffusion of a gas is inversely proportional to the square root of its molar mass. In other words, the lighter gas will effuse or diffuse more rapidly than the heavier gas.

Mathematically, Graham's law can be expressed as follows:

Rate of effusion or diffusion = Square root of (molar mass of gas2 / molar mass of gas1)

$$\frac{\text{rate}_1}{\text{rate}_2} = \sqrt{\frac{M_2}{M_1}}$$

Thomas Graham

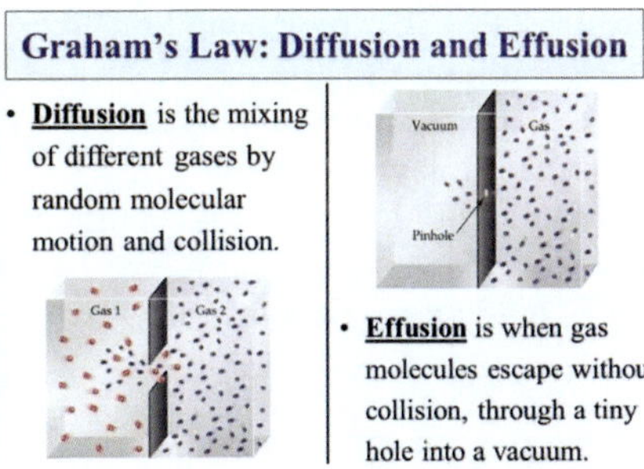

This law is important in many areas of science, particularly in the study of gas behavior and the properties of mixtures. It is commonly used in industrial processes such as gas separation, purification, and detection.

Applications of Graham's Law

Graham's law, also known as Graham's law of effusion or Graham's law of diffusion, is a principle in physical chemistry that describes the rate at which gases diffuse or effuse through a porous barrier. Here are some applications of Graham's law:

- **Separation of isotopes:** Graham's law is used in the separation of isotopes, which are atoms of the same element that have different numbers of neutrons. By taking advantage of the different rates at which isotopes diffuse or effuse through a porous barrier, researchers can separate isotopes for use in various applications, such as nuclear energy and medical imaging.
- **Understanding gas behavior:** Graham's law is used to understand the behavior of gases, such as their diffusion and effusion rates. By understanding these rates, researchers can predict the behavior of gases in various applications, such as in combustion engines and industrial processes.
- **Gas chromatography:** Graham's law is used in gas chromatography, which is a technique used to separate and analyze the components of a mixture. By taking advantage of the different rates at which the components of the mixture diffuse or effuse through a porous barrier, researchers can separate and identify the individual components of the mixture.
- **Study of atmospheric gases:** Graham's law is used to study the behavior of gases in the Earth's atmosphere, such as their diffusion and effusion rates. By understanding these rates, researchers can predict the behavior of atmospheric gases, such as their transport and distribution, and their impact on climate change.

Graham's law has many important applications in physical chemistry, environmental science, and engineering. It provides a fundamental framework for understanding the behavior of gases and has led to many important technological advancements.

62. GRAVITOMAGNETIC LORENTZ FORCE

The gravitomagnetic Lorentz force, also known as the Lense-Thirring effect, is a prediction of Einstein's theory of general relativity. It describes the effect of a rotating massive object on the motion of a test particle in its vicinity, similar to how a magnetic field affects the motion of a charged particle.

The gravitomagnetic Lorentz force arises due to the curvature of space-time around a rotating massive object, which causes a frame dragging effect. As a result, a test particle with mass m moving with velocity v in the vicinity of the rotating object experiences a force F given by:

$$F = (2G/c^2) \times (J \times v)$$

where G is the gravitational constant, c is the speed of light, J is the angular momentum of the rotating object, and × denotes the cross product.

The gravitomagnetic Lorentz force has important astrophysical applications, including the study of the behavior of matter and radiation around rotating black holes and neutron stars. It also has potential technological applications in the development of future space propulsion systems.

Josef Lense **Hans Thirring**

(https://www.researchgate.net/figure/Discoverers-of-the-frame-dragging-effect-in-general-relativity-Josef-Lense-left-and_fig13_226056893)
(https://geschichte.univie.ac.at/en/persons/hans-thirring)

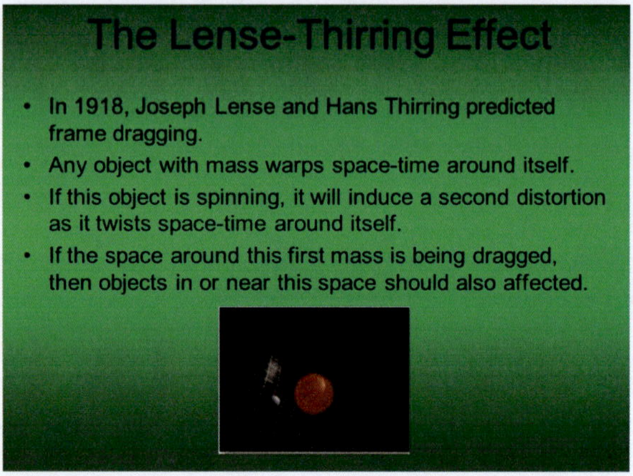

Applications of Gravitomagnetic Lorentz Force

The Gravitomagnetic Lorentz force, also known as the Lense-Thirring effect, is a phenomenon in general relativity where a spinning mass induces a frame-dragging effect on space-time. This effect has several practical applications in astrophysics and space technology, including:

♦ **Frame-dragging measurement:** The Lense-Thirring effect allows scientists to measure the spin of massive objects, such as black holes and neutron stars, by observing the precession of the orbits of other objects around them.

♦ **Satellite orbit design:** The frame-dragging effect can also be used to design satellite orbits that take advantage of the rotation of the Earth to save fuel and increase the lifetime of the satellite. These orbits are known as "gravity-assist" or "gravity-gradient" orbits.

♦ **Precision tests of general relativity:** The Lense-Thirring effect provides a precise prediction of the effect of a spinning mass on space-time, which can be used to test the predictions of general relativity.

♦ **Gravity-wave detection:** The Lense-Thirring effect is also a source of gravitational waves, which are ripples in the fabric of space-time. Detecting these waves is one of the primary goals of modern gravitational wave observatories, such as LIGO and VIRGO.

Gravitomagnetic Lorentz force has several practical applications in astrophysics and space technology, including frame-dragging measurement, satellite orbit design, precision tests of general relativity, and gravity-wave detection.

63. HAMILTON–JACOBI EQUATION

The Hamilton–Jacobi equation is a partial differential equation in classical mechanics, which is named after William Rowan Hamilton and Carl Gustav Jacob Jacobi. It provides an alternative formulation of classical mechanics, which is closely related to the principle of least action.

The Hamilton–Jacobi equation can be derived from the Hamiltonian formulation of classical mechanics, which is based on the concept of phase space. In this formulation, the state of a mechanical system is described by a point in phase space, which is a six-dimensional space consisting of the position and momentum coordinates of all particles in the system.

The Hamilton–Jacobi equation is a first-order partial differential equation that relates the Hamiltonian function of a mechanical system to a so-called "action" function. The action function is a scalar function that describes the motion of the system, and it is related to the path that the system takes through phase space.

The Hamilton–Jacobi equation can be used to solve for the motion of a mechanical system, by finding the action function that satisfies the equation. This is done by separating the variables in the Hamilton–Jacobi equation, and then solving for the action function in terms of the position and time coordinates of the system.

Once the action function is known, the trajectory of the system can be found by using a set of canonical equations, known as Hamilton's equations. These equations relate the position and momentum coordinates of the system to the partial derivatives of the action function with respect to these coordinates.

The Hamilton–Jacobi equation has many applications in classical mechanics, including the study of conservative systems, integrable systems, and chaotic systems. It also has important applications in quantum mechanics, where it is used to solve for the wave function of a quantum mechanical system.

$$H\left(q_k, \frac{\partial S}{\partial q_k}, t\right) + \frac{\partial S}{\partial t} = 0$$

William R Hamilton

Carl Gustav Jacobi

Applications of Hamilton–Jacobi Equation

Hamilton-Jacobi equation is an important equation in classical mechanics and has a number of applications, some of which are:

- **Celestial mechanics:** In celestial mechanics, the Hamilton-Jacobi equation is used to describe the motion of celestial bodies, such as planets, moons, asteroids, and comets. It is used to derive equations of motion, calculate orbits, and make predictions about the positions and movements of celestial bodies.
- **Quantum mechanics:** In quantum mechanics, the Hamilton-Jacobi equation is used to study the behavior of quantum particles. It is used to derive the Schrödinger equation and to solve it for simple systems. The Hamilton-Jacobi equation is used in the study of quantum field theory.
- **Optics:** In optics, the Hamilton-Jacobi equation is used to study the propagation of light through different media. It is used to derive the laws of reflection and refraction, and to calculate the paths of rays of light.
- **Control theory:** In control theory, the Hamilton-Jacobi equation is used to study the behavior of control systems. It is used to derive optimal control laws in a variety of applications, such as robotics, aerospace engineering, and industrial control.
- **Fluid dynamics:** In fluid dynamics, the Hamilton-Jacobi equation is used to study the behavior of fluids. It is used to derive the equations of motion for fluids, and to study the behavior of vortices, waves, and other phenomena.
- **Classical mechanics:** In classical mechanics, the Hamilton-Jacobi equation is used to study the motion of particles and systems. It is used to derive equations of motion, calculate trajectories, and make predictions about the behavior of mechanical systems.

64. HAMILTON'S EQUATIONS

Hamilton's equations are a set of canonical equations that describe the evolution of a Hamiltonian system, which is a system that has a Hamiltonian function that describes its energy. The equations were first developed by Irish mathematician William Rowan Hamilton in the 19th century.

Hamilton's equations can be written in terms of the Hamiltonian function H(q,p) of a system, where q and p are the generalized coordinates and momenta of the system, respectively. The equations are:

$$dq/dt = \partial H/\partial p$$

$$dp/dt = -\,\partial H/\partial q$$

These equations describe how the generalized coordinates and momenta of a system evolve over time. They are often used to describe the motion of particles in classical mechanics, but they can also be used in other fields such as quantum mechanics, statistical mechanics, and optics.

Hamilton's equations are important because they provide a powerful mathematical tool for studying the dynamics of physical systems. They can be used to derive the equations of motion of a system and to determine the stability and behavior of a system over time.

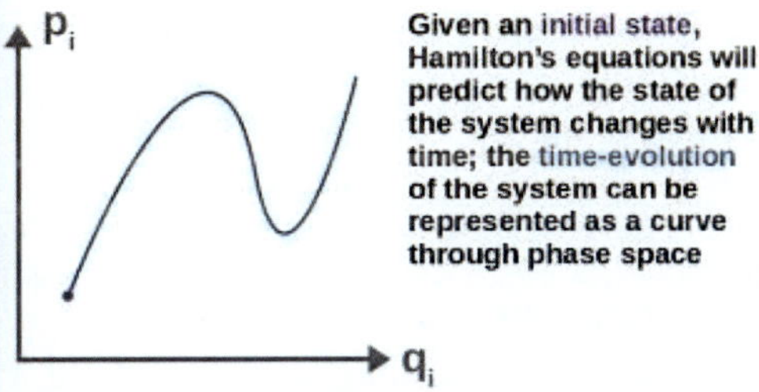

William Rowan Hamilton
(Credit: Royal Irish Academy)

Applications of Hamilton's Equations

Hamilton's equations have many applications in physics, mechanics, and engineering, and are an important tool for understanding the behavior of complex systems. Some of the applications of Hamilton's equations are:

- **Classical mechanics:** Hamilton's equations provide an alternative formulation of classical mechanics. They are used to find the equations of motion of a system described by a Hamiltonian, which is a function of the system's position and momentum.
- **Celestial mechanics:** Hamilton's equations are used to study the motion of celestial bodies, such as planets and moons. They can be used to find the equations of motion of a system consisting of two or more celestial bodies, such as the Earth and the Moon.
- **Quantum mechanics:** Hamilton's equations can be used to derive the Schrödinger equation, which is the fundamental equation of quantum mechanics. The Schrödinger equation describes the time evolution of a quantum mechanical system.
- **Optics:** Hamilton's equations are used in optics to study the propagation of light in materials. They can be used to find the equations of motion of the electric and magnetic fields that describe the behavior of light in a material.
- **Control theory:** Hamilton's equations are used in control theory to find the optimal control of a system. They can be used to find the control that minimizes a cost function, subject to constraints on the control and the state of the system.
- **Robotics:** Hamilton's equations are used in robotics to model the dynamics of robots. They can be used to find the equations of motion of a robot, and to design controllers that control the motion of the robot.
- **Fluid dynamics:** Hamilton's equations can be used in fluid dynamics to study the motion of fluids. They can be used to find the equations of motion of a fluid, and to study the stability of fluid flows.
- **Statistical mechanics:** Hamilton's equations can be used in statistical mechanics to study the behavior of large systems of particles. They can be used to derive the equations of motion of the particles, and to calculate the thermodynamic properties of the system.
- **Chaos theory:** Hamilton's equations are used in chaos theory to study the behavior of dynamical systems that are highly sensitive to initial conditions. They can be used to analyze the stability and predictability of such systems.

65. Hammond–Leffler Postulate

The Hammond–Leffler postulate is a concept in organic chemistry that predicts the transition state of a reaction. It states that the structure of the transition state of a reaction resembles the structure of the species, either the reactant or product, which it is closest to in energy. This means that if the reactants of a reaction are energetically closer to the transition state, then the transition state will resemble the reactants more than the products. Conversely, if the products are closer in energy to the transition state, then the transition state will resemble the products more than the reactants. The postulate is named after its proponents, George Hammond and Paul Leffler.

George Hammond

(Credit:https://commons.wikimedia.org/wiki/File:GeorgeSHammond.png)

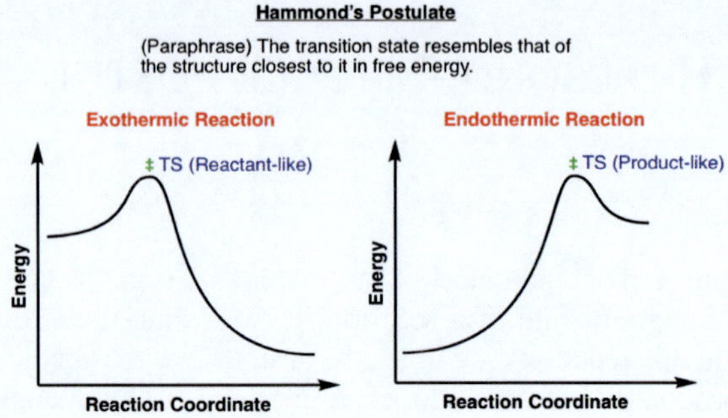

Applications of Hammond–Leffler Postulate

The Hammond-Leffler postulate, also known as the principle of microscopic reversibility, is a fundamental concept in chemical kinetics that states that the transition state for a chemical reaction resembles either the reactants or the products, depending on which is lower in energy. This principle has a wide range of applications in chemistry, including:

- **Reaction mechanisms:** The Hammond-Leffler postulate can be used to predict the relative rates of different steps in a reaction mechanism. If the transition state for a particular step resembles the reactants, that step will be slower than a step in which the transition state resembles the products.
- **Solvent effects:** The Hammond-Leffler postulate can be used to explain the effect of solvents on reaction rates. Solvents can stabilize either the reactants or the products more than the transition state, which can change the relative rates of different steps in a reaction.
- **Enzyme catalysis:** The Hammond-Leffler postulate can be applied to enzyme catalysis to predict the rate-determining step and the effect of mutations or changes in the enzyme structure on the rate of the reaction.
- **Photochemistry:** The Hammond-Leffler postulate can be used to predict the products of photochemical reactions, as the transition state for a photochemical reaction often resembles the reactants more than the products.
- **Chemical Reactivity:** The Hammond-Leffler postulate can be used to explain the selectivity of reactions, particularly in cases where there are competing reaction pathways that lead to different products. By considering the relative energies of the transition states, it is possible to predict which pathway is more likely to occur.

66. HEISENBERG UNCERTAINTY PRINCIPLE

The Heisenberg Uncertainty Principle is a fundamental principle in quantum mechanics that states that it is impossible to simultaneously know the exact position and momentum of a particle. This principle was first proposed by Werner Heisenberg in 1927 and is one of the most important principles in quantum mechanics. The uncertainty principle arises because the act of measuring one property of a particle (such as its position) disturbs the other property (such as its momentum).

Mathematically, the Heisenberg Uncertainty Principle can be expressed as

$$\Delta x \, \Delta p \geq h / 4\pi$$

where, Δx is the uncertainty in the position of the particle, Δp is the uncertainty in the momentum of the particle, and h is Planck's constant. This means that the product of the uncertainties in position and momentum of a particle must always be greater than or equal to a certain minimum value.

Werner Heisenberg

(Credit:https://www.bild.bundesarchiv.de/dba/de/search/?query=Bild+183 -R57262)

Heisenberg Uncertainty Principle

According to Heisenberg's Uncertainty principle, " **Position and momentum of an electron cannot be determined simultaneously with absolute accuracy.**"

Atomic Structure

Equation

$$\Delta x \,.\, \Delta p \geq \frac{h}{4\pi}$$

$$\Delta x \,.\, m\,\Delta v \geq \frac{h}{4\pi}$$

$$\Delta x \,.\, \Delta v \geq \frac{h}{4\pi m}$$

Applications of Heisenberg Uncertainty Principle

The Heisenberg Uncertainty Principle has important implications in many areas of physics, including atomic and nuclear physics, quantum computing, and quantum field theory. It places fundamental limits on the precision with which certain measurements can be made and has led to a deeper understanding of the nature of matter and the universe. It helps us understand the behavior of subatomic particles.

- **Atomic and molecular physics:** The Heisenberg Uncertainty Principle is used to understand the behavior of electrons in atoms and molecules. It helps explain why electrons can only occupy certain energy levels in atoms and why atoms and molecules have discrete spectra.
- **Quantum mechanics:** The Heisenberg Uncertainty Principle is a fundamental principle in quantum mechanics. It places limits on the precision with which certain measurements can be made and is used to understand the behavior of subatomic particles.
- **Quantum computing:** The Heisenberg Uncertainty Principle is important in the development of quantum computing. The principle sets a limit on the precision with which certain measurements can be made, which affects the accuracy of quantum computing operations.
- **Quantum field theory:** The Heisenberg Uncertainty Principle is used in quantum field theory to understand the behavior of quantum fields. It helps explain why particles can appear and disappear spontaneously and why the vacuum of space is not empty but is filled with virtual particles.
- **Uncertainty in metrology:** The Heisenberg Uncertainty Principle has implications for the precision of measurements in metrology, which is the science of measurement. It places limits on the precision with which certain measurements can be made and is used to understand the limitations of modern measurement techniques.

67. HELLMANN–FEYNMAN THEOREM

The Hellmann-Feynman theorem is a fundamental result in quantum mechanics that relates the derivative of the total energy of a system with respect to a parameter to the expectation value of the derivative of the Hamiltonian operator with respect to the same parameter.

The Hamiltonian operator is a mathematical operator that describes the total energy of a quantum mechanical system. The theorem states that if the Hamiltonian operator depends on a parameter λ, then the derivative of the total energy of the system with respect to λ is equal to the expectation value of the derivative of the Hamiltonian with respect to λ:

$$\mathbf{dE/d\lambda = \langle \partial H/\partial \lambda \rangle}$$

Here, E is the total energy of the system, H is the Hamiltonian operator, λ is the parameter, and <...> denotes the expectation value.

The Hellmann-Feynman theorem has important applications in quantum chemistry and solid state physics, where it is used to calculate properties such as molecular and crystal structures, chemical reactivity, and electronic properties of materials. It provides a way to compute the response of a quantum mechanical system to an external perturbation without explicitly solving the Schrödinger equation.

Richard Feynman

Hans Gustav Hellmann

(Credit:https://commons.wikimedia.org/wiki/File:Richard_feynman_-_fermilab1.jpg)

(Credit:https://www.archiv.ub.uni-stuttgart.de/alumni/hellmann_hans.html)

Applications of Hellmann–Feynman Theorem

The Hellmann-Feynman theorem is a principle in quantum mechanics that relates the change in energy of a system to the change in potential energy due to the displacement of its constituent particles. Here are some applications of the Hellmann-Feynman theorem:

- **Determining atomic and molecular properties:** The Hellmann-Feynman theorem can be used to determine the properties of atoms and molecules, such as their ionization energies, electron affinities, and bond lengths. By understanding the relationship between the energy and potential energy of the system, researchers can calculate these properties accurately.
- **Designing materials:** The Hellmann-Feynman theorem is used in the design of materials, such as semiconductors and superconductors. By understanding the relationship between the energy and potential energy of the system, researchers can design materials with specific properties, such as high conductivity or low thermal resistance.
- **Studying chemical reactions:** The Hellmann-Feynman theorem is used in the study of chemical reactions, such as those that occur in catalysis and photosynthesis. By understanding the relationship between the energy and potential energy of the system, researchers can predict the outcome of chemical reactions and design more efficient and effective catalysts.
- **Calculating forces:** The Hellmann-Feynman theorem is used to calculate forces in systems, such as the forces between atoms in a molecule or the forces between molecules in a material. By understanding the relationship between the energy and potential energy of the system, researchers can calculate these forces accurately.

Hellmann-Feynman theorem has many important applications in quantum mechanics, materials science, and chemistry. It provides a fundamental framework for understanding the behavior of systems and has led to many important technological advancements.

68. HELMHOLTZ EQUATION

The Helmholtz equation is a partial differential equation that arises in many areas of physics, including electromagnetism, acoustics, and fluid mechanics. The equation is named after the German physicist Hermann von Helmholtz, who first derived it in the 19th century.

The Helmholtz equation is given by:

$$\nabla^2\varphi + k^2\varphi = 0$$

where φ is a scalar function, ∇^2 is the Laplacian operator, k is the wave number, and the equation is usually solved in the context of some boundary conditions.

The Helmholtz equation describes the behavior of waves in a medium and is used to model a wide range of physical phenomena, including the propagation of electromagnetic waves, the behavior of acoustic waves in fluids, and the behavior of wave functions in quantum mechanics.

Hermann von Helmholtz

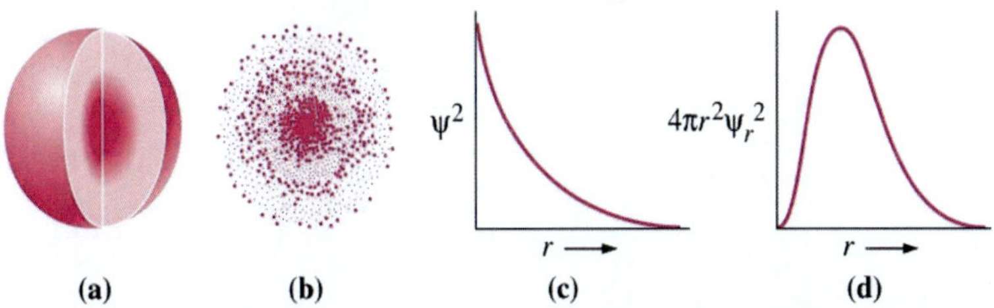

Applications of Helmholtz Equation

The Helmholtz equation has a wide range of applications in physics, engineering, and other fields. Here are a few specific examples:

- **Electromagnetic wave propagation:** The Helmholtz equation is used to describe the propagation of electromagnetic waves in free space and in materials. It is used in the design of antennas, waveguides, and other components of communication systems.
- **Fluid mechanics:** The Helmholtz equation is used to describe the behavior of fluids, including the flow of air and water. It is used in the design of aircraft, ships, and other vehicles that operate in fluid environments.
- **Quantum mechanics:** The Helmholtz equation is used in quantum mechanics to describe the behavior of wave functions in free space and in potential wells. It is used to model the behavior of electrons in semiconductors and other materials. This has applications in fields such as solid-state physics, semiconductor device design, and quantum computing.
- **Antenna design:** The Helmholtz equation is used to model the behavior of electromagnetic waves in free space and in materials, which is essential for the design of antennas. Antennas are used in communication systems, radar systems, and other applications that require the transmission and reception of electromagnetic waves.
- **Acoustic engineering:** The Helmholtz equation is used to model the behavior of sound waves in air, water, and other fluids, which is essential for the design of acoustic devices such as speakers, microphones, and noise-canceling systems.
- **Medical imaging:** The Helmholtz equation is used in medical imaging techniques such as magnetic resonance imaging (MRI) and ultrasound. These techniques rely on the propagation and detection of electromagnetic or acoustic waves through the body, which can be modeled using the Helmholtz equation.

Helmholtz equation is a fundamental equation in physics that has many applications in a wide range of fields, including electromagnetism, acoustics, fluid mechanics, and quantum mechanics.

69. HELMHOLTZ FREE ENERGY

Helmholtz free energy, also known as Helmholtz energy, is a thermodynamic function that is used to describe the amount of work that can be extracted from a system at constant temperature and volume. It is named after the German physicist Hermann von Helmholtz, who introduced the concept in the mid-19th century.

The Helmholtz free energy is defined as:

$$F = U - TS$$

where F is the Helmholtz free energy, U is the internal energy of the system, T is the absolute temperature, and S is the entropy of the system.

The Helmholtz free energy is a state function, which means that its value depends only on the current state of the system and not on the path taken to reach that state. It is related to other thermodynamic functions such as the enthalpy, entropy, and Gibbs free energy, and can be used to calculate the maximum amount of work that can be extracted from a system under specific conditions.

Hermann von Helmholtz

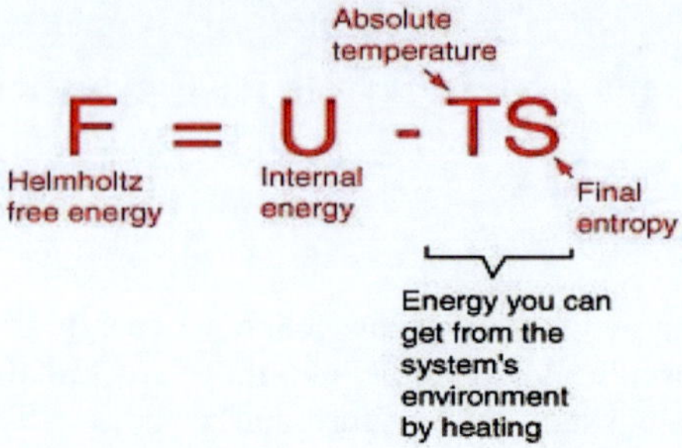

Applications of Helmholtz Free Energy

Applications of the Helmholtz free energy include:

- **Chemical reactions:** The Helmholtz free energy is used to calculate the maximum amount of work that can be obtained from a chemical reaction at constant temperature and volume. This information can be used to predict whether a reaction will occur spontaneously or not.
- **Materials science:** The Helmholtz free energy is used to describe the thermodynamic properties of materials, such as their heat capacity, phase transitions, and stability. This information is important for the design and development of new materials for various applications.
- **Thermodynamics:** The Helmholtz free energy is a fundamental concept in thermodynamics and is used to describe the behavior of many different systems, including gases, liquids, and solids.
- **Biological systems:** The Helmholtz free energy is used in the study of biological systems, such as enzymes and proteins. It provides information about the stability of these systems and the energy required for certain processes to occur, such as the folding and unfolding of proteins.
- **Electrochemistry:** The Helmholtz free energy is used to describe the behavior of electrochemical systems, such as batteries and fuel cells. It provides information about the maximum amount of work that can be obtained from these systems and the efficiency of the energy conversion process.

Helmholtz free energy is an important thermodynamic function that has many applications in chemistry, materials science, and other fields. It provides valuable information about the maximum amount of work that can be extracted from a system at constant temperature and volume, which is essential for understanding the behavior of many different physical and chemical systems.

70. Henry's Law

Henry's law is a principle in physical chemistry that describes the solubility of a gas in a liquid at a constant temperature. It states that the amount of gas dissolved in a liquid is proportional to the partial pressure of the gas above the liquid.

In mathematical terms, Henry's law can be expressed as:

$$C = k_H \times P$$

where C is the concentration of the gas in the liquid, P is the partial pressure of the gas above the liquid, and k_H is the Henry's law constant, which depends on the temperature, the solvent, and the gas. The law is valid for dilute solutions and for gases that do not react chemically with the solvent.

Henry's law has many practical applications, including in the fields of environmental chemistry, oceanography, and chemical engineering. For example, it is used to describe the absorption of gases in the blood, the release of gases from water in aquatic ecosystems, and the separation of gases in industrial processes.

William Henry

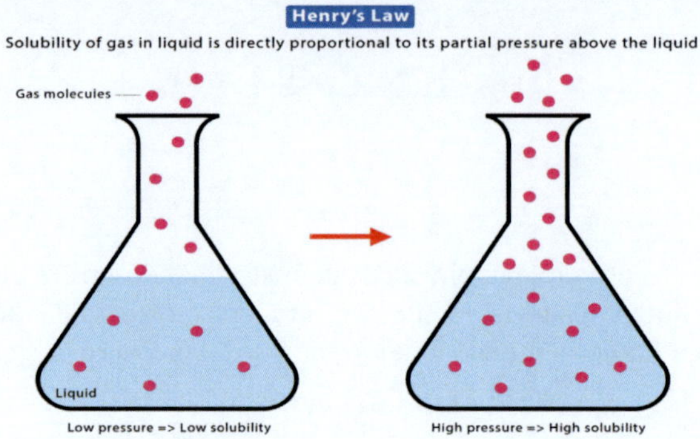

Applications of Henry's Law

Henry's law is a principle in physical chemistry that describes the behavior of gases in solution. Here are some applications of Henry's law:

♦ **Carbonation of beverages:** Henry's law is used to explain the carbonation of beverages, such as soda and beer. By understanding how the partial pressure of carbon dioxide above the liquid affects its solubility in the beverage, manufacturers can control the carbonation level to produce the desired taste and texture.

♦ **Purification of gases:** Henry's law is used in the purification of gases, such as the removal of carbon dioxide from natural gas or the separation of oxygen and nitrogen from air. By controlling the partial pressure of the gas above the liquid, the desired gas can be selectively dissolved, and the impurities can be removed.

♦ **Understanding gas exchange in aquatic systems:** Henry's law is used to understand gas exchange in aquatic systems, such as the transfer of oxygen and carbon dioxide between the atmosphere and bodies of water. By understanding the relationship between the partial pressure of gases in the atmosphere and their solubility in water, researchers can predict the dissolved gas concentrations in aquatic systems.

♦ **Chemical analysis:** Henry's law is used in chemical analysis techniques, such as gas chromatography and mass spectrometry. By understanding the solubility of gases in liquids, researchers can extract and analyze gas samples from various sources, such as blood or air.

Henry's law has many important applications in physical chemistry, environmental science, and engineering. It provides a fundamental framework for understanding the behavior of gases in solution and has led to many important technological advancements.

71. HERTZ OBSERVATIONS

Hertz observations refer to the experimental observations made by the German physicist Heinrich Hertz in the late 19th century regarding the properties of electromagnetic waves. Hertz was the first person to demonstrate the existence of electromagnetic waves, which had been predicted by James Clerk Maxwell's theory of electromagnetism.

Hertz conducted a series of experiments in which he generated and detected electromagnetic waves using oscillating circuits. He observed that these waves had many properties similar to those of light, such as reflection, refraction, diffraction, and polarization. He also showed that the velocity of these waves was equal to the speed of light.

Hertz's observations confirmed Maxwell's theory and provided experimental evidence for the wave nature of electromagnetic radiation. They laid the foundation for the development of wireless communication and other technologies that rely on the transmission and reception of electromagnetic waves.

Heinrich Hertz

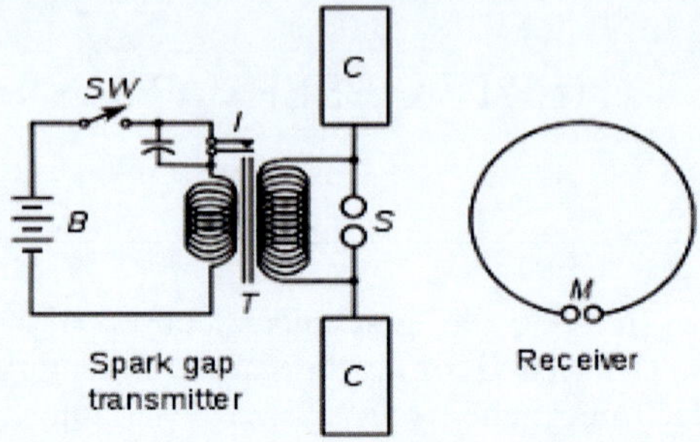

Applications of Hertz Observations

Hertz's observations and discoveries have had a wide range of applications in modern physics, technology, and engineering. They have helped to shape our understanding of the universe and our ability to control and manipulate it, and have led to many important technological advancements that have transformed the way we live and work. Here are some applications of Hertz's observations:

- **Development of wireless communication:** Hertz's experiments demonstrated the existence of electromagnetic waves and their ability to travel through space. This paved the way for the development of wireless communication technologies, such as radio and television broadcasting, satellite communications, and mobile phones.
- **Understanding of light:** Hertz's experiments helped to confirm that light is an electromagnetic wave, as predicted by James Clerk Maxwell's theory. This led to a better understanding of the nature of light and its properties, which had important applications in fields such as optics, spectroscopy, and photovoltaics.
- **Development of radar:** Hertz's discovery of the reflection and refraction of electromagnetic waves led to the development of radar, which is used in a wide range of applications, such as aviation, meteorology, and military surveillance.
- **Fundamental research in electromagnetism:** Hertz's experiments contributed to the development of the field of electromagnetism, which has had a profound impact on our understanding of the universe and our ability to control and manipulate it. His work helped to advance the understanding of electric and magnetic fields, and contributed to the development of modern technologies such as electric motors, generators, and transformers.

72. HESS'S LAW

Hess's law, also known as Hess's law of constant heat summation, is a fundamental principle in thermodynamics that states that the enthalpy change of a chemical reaction is independent of the pathway between the initial and final states of the system. In other words, the change in enthalpy of a chemical reaction is the same whether the reaction occurs in one step or in several steps.

Enthalpy is a thermodynamic property that describes the heat content of a system. It is denoted by the symbol H and is measured in units of energy per unit of substance, such as joules per mole. The enthalpy change of a chemical reaction is the difference between the enthalpies of the reactants and products, and is usually expressed as ΔH.

Hess's law is based on the principle of conservation of energy, which states that energy cannot be created or destroyed, only transferred from one form to another. It is named after the Swiss-born Russian chemist Germain Hess, who first formulated it in 1840.

Hess's law has many practical applications, including in the field of chemical engineering, where it is used to design and optimize chemical processes. It is also used in analytical chemistry to calculate the enthalpy changes of reactions that are difficult to measure directly.

Germain Hess

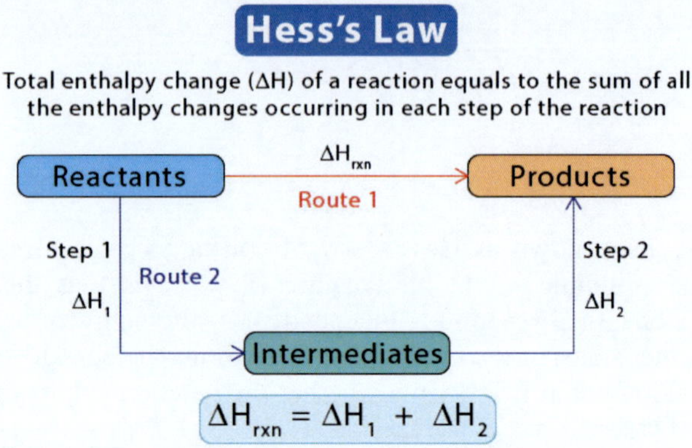

Applications of Hess's Law

Hess's law is a fundamental principle in thermodynamics that has many important applications in chemistry and materials science. It provides a powerful tool for understanding the energy changes associated with chemical reactions and physical processes, and has led to many important technological advancements. This principle has many applications in chemistry, including:

- **Determination of heats of reaction:** Hess's law is used to determine the enthalpy change of a chemical reaction that cannot be measured directly. This is done by using a series of reactions whose enthalpies of reaction are known, and applying Hess's law to calculate the enthalpy change of the desired reaction.
- **Calculation of combustion energies:** Hess's law is used to calculate the combustion energies of various fuels, such as coal, oil, and natural gas. This is important for understanding the energy content of these fuels, and for predicting their combustion products.
- **Design of energy storage systems:** Hess's law is used in the design of energy storage systems, such as batteries and fuel cells. By understanding the enthalpy changes involved in the charging and discharging of these systems, engineers can optimize their design and performance.
- **Prediction of phase transitions:** Hess's law is used to predict the enthalpy changes associated with phase transitions, such as melting and boiling. This is important for understanding the behavior of materials under different conditions, and for designing processes that involve phase transitions.

73. HOHENBERG–KOHN THEOREM

The Hohenberg-Kohn theorem is a fundamental result in the field of density functional theory (DFT), a widely used method in quantum mechanics to calculate the electronic structure of molecules and solids. The theorem states that the ground-state electron density of a system uniquely determines the external potential acting on the electrons, up to an arbitrary constant. In other words, the total energy of a system is a functional of the electron density.

The Hohenberg-Kohn theorem was first proposed by Pierre Hohenberg and Walter Kohn in 1964, and it provided a solid theoretical foundation for DFT. It allows for the efficient calculation of electronic properties of materials, such as band structures, vibrational frequencies, and chemical reaction rates, with relatively simple computer codes.

The Hohenberg-Kohn theorem has been widely used in materials science, chemistry, and physics to study a variety of systems, ranging from small molecules to complex materials. It has also inspired the development of many extensions and modifications of DFT to address the limitations of the original theorem, such as the treatment of excited states and magnetic systems.

Pierre Hohenberg **Walter Kohn**

(Credit: https://link.springer.com/article/10.1007/s10955-019-02285-5)
(Credit: https://commons.wikimedia.org/wiki/File:Walter_Kohn.jpg)

Hohenberg-Kohn theorem

The total energy of an interacting inhomogeneous electron gas in the presence of an external potential $V_{ext}(r)$ is a functional of the density ρ

$$E = \int V_{ext}(\vec{r})\rho(\vec{r})d\vec{r} + F[\rho]$$

Applications of Hohenberg–Kohn Theorem

The Hohenberg-Kohn theorem is a fundamental result in the field of density functional theory (DFT), a method for calculating the electronic structure of materials. The theorem has many applications in materials science and computational chemistry, including:

- **Prediction of material properties:** The Hohenberg-Kohn theorem provides a theoretical framework for predicting the properties of materials based on their electronic structure. This includes properties such as band gaps, dielectric constants, and magnetic moments, which are important for understanding the behavior of materials in various applications.
- **Design of new materials:** DFT calculations based on the Hohenberg-Kohn theorem can be used to design new materials with specific properties. By predicting the electronic structure of a material, researchers can identify promising candidates for various applications, such as catalysis, energy storage, and electronic devices.
- **Understanding of surface chemistry:** The Hohenberg-Kohn theorem provides a theoretical framework for understanding the electronic structure of surfaces and interfaces, which is important for designing materials for applications such as sensors and catalysts.
- **Development of new computational methods:** The Hohenberg-Kohn theorem has inspired the development of new computational methods for solving the electronic structure problem, such as time-dependent DFT and hybrid functionals. These methods have enabled more accurate and efficient calculations of the electronic structure of materials.

Hohenberg-Kohn theorem is a powerful tool for understanding the electronic structure of materials, and has many applications in materials science and computational chemistry. It provides a theoretical foundation for predicting material properties, designing new materials, and developing new computational methods for solving the electronic structure problem.

74. HOOKE'S LAW

Hooke's Law is a fundamental principle in physics that describes the behavior of an ideal spring. It states that the force exerted by a spring is directly proportional to its extension or compression from its equilibrium position, provided the deformation is within the spring's elastic limit. This means that if a spring is extended or compressed by a certain amount, it will exert a force that is proportional to that amount.

Mathematically, Hooke's Law can be expressed as

$$F = -k\,x$$

where F is the force exerted by the spring, x is the displacement from the equilibrium position, and k is the spring constant, which is a measure of the stiffness of the spring.

Robert Hooke

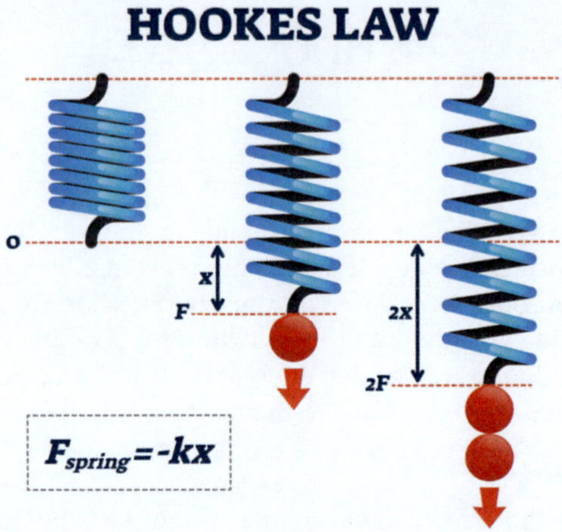

(Vanstone, Emma. 2022. "What is Hooke's Law?". Accessed April 20, 2022)

Applications of Hooke's Law

Hooke's Law has many applications in physics and engineering, including:

- **Mechanics:** Hooke's Law is used to describe the behavior of springs in mechanical systems, such as in suspension systems in vehicles, shock absorbers, and mechanical watches.
- **Material science:** Hooke's Law is used to characterize the elasticity of materials, which is a measure of their ability to deform and return to their original shape when a force is applied.
- **Geophysics:** Hooke's Law is used to describe the behavior of rocks and soils in geophysics. It helps explain how rocks and soils deform and respond to stress and strain.
- **Electrical engineering:** Hooke's Law is used to describe the behavior of electrical circuits, such as in the design of strain gauges, which are used to measure mechanical stress.

Hooke's Law is a fundamental principle in physics that describes the behavior of springs and other elastic materials. It has numerous applications in physics and engineering, and it is an essential tool for understanding the behavior of mechanical systems and materials.

75. Hopkinson's Law

Hopkinson's law, also known as the Hopkinson pressure bar technique, is a method for measuring the dynamic properties of materials, such as the strength and deformation characteristics, at high strain rates. The method involves subjecting a specimen to a rapid compression or tension load using a split Hopkinson pressure bar apparatus, and measuring the resulting stress and strain waveforms using high-speed sensors.

The technique was first developed by the British engineer John Hopkinson in the late 19th century, and has since become a widely used tool in materials science, mechanical engineering, and structural design. Hopkinson's law describes the relationship between the stress and strain waveforms measured during a compression or tension test, and the material properties, such as the Young's modulus, Poisson's ratio, and yield strength.

Hopkinson's law is particularly useful for studying the behavior of materials at high strain rates, which are encountered in many real-world applications, such as impact and explosion dynamics, crashworthiness of structures, and ballistics. The technique has also been used to investigate the response of materials to extreme conditions, such as high temperature and pressure, and to develop new materials with improved properties.

John Hopkinson

(https://commons.wikimedia.org/wiki/File:John_Hopkinson_%282%29.jpg)

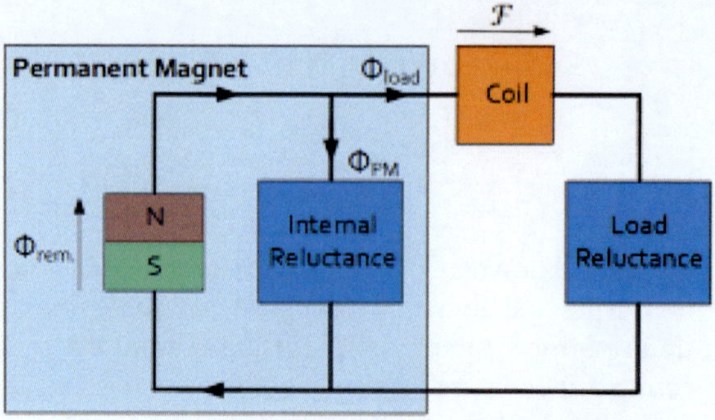

Applications of Hopkinson's Law

Hopkinson's law, also known as the reciprocal relationship between fatigue life and strength, is a fundamental principle in the study of materials science and engineering. The law has many important applications in the design of components, optimization of manufacturing processes, prediction of material behavior, and development of new materials.

- **Design of high-stress components:** Hopkinson's law is used to design components that are subjected to high stress or fatigue. By understanding the relationship between fatigue life and strength, engineers can design components that are able to withstand repeated loading without failure.

- **Optimization of manufacturing processes:** Hopkinson's law is used to optimize manufacturing processes for materials. By understanding the relationship between strength and fatigue life, engineers can adjust the manufacturing process to produce materials with the desired properties.

- **Prediction of material behavior:** Hopkinson's law can be used to predict the behavior of materials under different loading conditions. By understanding the relationship between strength and fatigue life, researchers can predict how a material will behave under various stress and fatigue conditions, which is important for designing materials for different applications.

- **Development of new materials:** Hopkinson's law is used to develop new materials with improved fatigue resistance. By understanding the relationship between strength and fatigue life, researchers can develop new materials with improved properties that are able to withstand repeated loading without failure.

76. HUBBLE'S LAW

Hubble's law is a fundamental principle in astrophysics that describes the relationship between the distance of a galaxy and its apparent recession velocity. It is named after the astronomer Edwin Hubble who first discovered this phenomenon.

Hubble's law states that the farther away a galaxy is from us, the faster it is moving away from us. The velocity at which a galaxy is moving away from us is called its recession velocity. The recession velocity is proportional to the distance between us and the galaxy, meaning that the farther away a galaxy is, the faster it is moving away from us.

This relationship is described by the Hubble constant, which is a measure of the rate at which the universe is expanding. The current best estimate for the value of the Hubble constant is around 73 kilometers per second per megaparsec, meaning that for every megaparsec (about 3.26 million light-years) of distance, galaxies are moving away from us at a speed of 73 kilometers per second faster.

Hubble's law is a crucial piece of evidence supporting the Big Bang theory, as it indicates that the universe is expanding uniformly in all directions. It also provides a means to estimate the distances to faraway galaxies, which is essential for understanding the large-scale structure and evolution of the universe.

Edwin Hubble

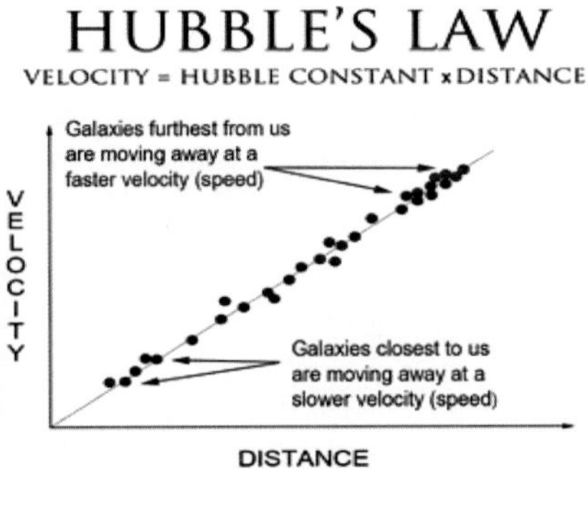

Applications of Hubble's Law

Hubble's law has played a vital role in shaping our understanding of the universe and continues to be a crucial tool for astronomers and cosmologists. Hubble's law has numerous applications in astrophysics and cosmology, including:

- **Determining the distance to galaxies:** Hubble's law provides a way to estimate the distance to faraway galaxies. By measuring the recession velocity of a galaxy and using the Hubble constant, astronomers can calculate its distance from us. This is known as the cosmological distance ladder and is used to determine the scale of the universe.
- **Measuring the expansion rate of the universe:** The Hubble constant provides a measure of the rate at which the universe is expanding. By measuring the recession velocity of galaxies at different distances, astronomers can determine the Hubble constant and track the evolution of the universe.
- **Studying the large-scale structure of the universe:** Hubble's law provides a means to study the distribution of galaxies in the universe. By mapping out the distribution of galaxies and measuring their recession velocities, astronomers can study the large-scale structure of the universe and the clustering of galaxies.
- **Testing cosmological models:** Hubble's law is a critical test of cosmological models. By comparing the predicted recession velocities of galaxies from different models to observed data, astronomers can determine which models best fit the data and refine our understanding of the universe's evolution.
- **Discovering new astronomical objects:** Hubble's law has led to the discovery of new astronomical objects such as quasars, which were first identified as bright, point-like sources with high redshifts (indicating they were at large distances from us).

77. HUND'S RULES

Hund's rules are a set of empirical rules used to predict the electronic configurations of atoms in their ground state. The rules were proposed by the German physicist Friedrich Hund in 1925 and are based on observations of the behavior of electrons in atoms. Hund's rules are:

- The lowest-energy configuration of an atom has the maximum number of unpaired electrons in a set of degenerate orbitals.
- When there are multiple orbitals of the same energy level available, electrons will occupy these orbitals singly before pairing up.
- When electrons pair up in the same orbital, they will have opposite spins.

Hund's rules help to explain the observed behavior of atoms in their ground state and are useful for predicting the electronic configurations of atoms and ions. They are widely used in chemistry and materials science to understand the properties of materials and chemical reactions.

The rules have also been extended to include more complex systems, such as molecules and solids, and have played an important role in the development of quantum mechanics and modern materials science.

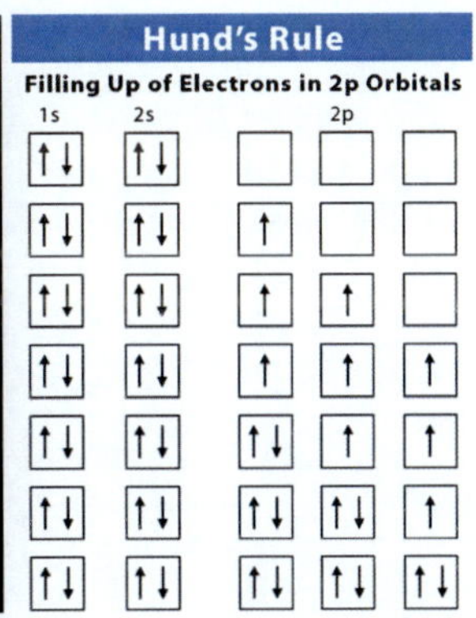

Friedrich Hund

(Credit:https://commons.wikimedia.org/wiki/File:Hund,Friedrich_192x_G%C3%B6ttingen.jpg)

Applications of Hund's Rules

Hund's rules are a set of principles that describe the distribution of electrons in atomic orbitals. They have a number of important applications, including:

- **Predicting the electronic configuration of atoms:** Hund's rules can be used to predict the electronic configuration of atoms, which is important for understanding their chemical properties. By following the rules, researchers can determine how many electrons are in each orbital and how they are distributed.
- **Understanding the magnetic properties of materials:** Hund's rules are important for understanding the magnetic properties of materials. For example, they can be used to predict the magnetic moment of a material, which is important for designing magnetic materials for various applications.
- **Predicting the properties of transition metal compounds:** Hund's rules are particularly important for predicting the properties of transition metal compounds, which are often complex and have a variety of electronic configurations. By understanding the rules, researchers can predict the magnetic properties, optical properties, and reactivity of these compounds.
- **Designing catalysts:** Hund's rules can be used to design catalysts for various chemical reactions. For example, by understanding the electronic configuration of the catalyst, researchers can predict its reactivity and selectivity for specific reactions.

Hund's rules are a fundamental principle in the study of atomic and molecular properties, and have many important applications in materials science, chemistry, and catalysis. They provide a framework for understanding how electrons are distributed in atoms and molecules, and can be used to predict the properties of materials and design new catalysts for various applications.

78. HUYGENS–FRESNEL PRINCIPLE

The Huygens-Fresnel principle is a fundamental principle in optics that explains how light waves propagate and interfere with each other. The principle is named after the Dutch physicist Christiaan Huygens and the French physicist Augustin-Jean Fresnel.

Mathematically, Huygens' principle can be expressed as follows: "Each point on a wave-front can be considered as a new source of waves, which propagate with the same speed and frequency as the original wave. The envelope of all these waves gives the position of the new wave-front at a later time."

The Huygens-Fresnel principle states that every point on a wave-front can be considered as a source of secondary spherical waves, called Huygens-Fresnel wavelets, which propagate in all directions with a velocity equal to the velocity of the original wave. The sum of all these wavelets at a given point and time determines the amplitude and phase of the resulting wave at that point and time.

The Huygens-Fresnel principle is fundamental to many areas of optics and has had a significant impact on the development of modern optical technology, including the design of lenses, mirrors, and other optical components.

Christiaan Huygens **Augustin-Jean Fresnel**

(Painting by Caspar Netscher, Museum Boerhaave, Leiden, https://flickr.com/photos/38483926@N07/3757963275)

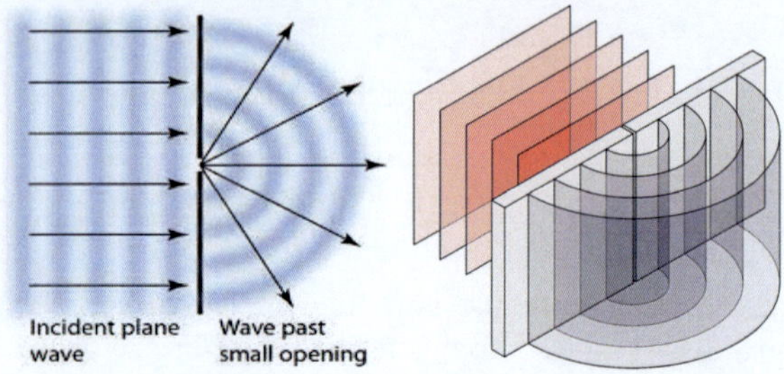

Applications of Huygens–Fresnel Principle

The Huygens-Fresnel principle is a fundamental principle in the study of optics, which describes how light waves propagate through space. It has many important applications, including:

- **Understanding the diffraction of light:** The Huygens-Fresnel principle can be used to understand the diffraction of light, which is the bending and spreading of light waves as they pass through narrow openings or around obstacles. By understanding how the wavefronts of light propagate, researchers can predict the diffraction patterns that result.
- **Reflection and refraction:** Huygens' principle is used to explain the behavior of waves at boundaries between media with different refractive indices. According to Huygens' principle, when a wave-front encounters a boundary, new secondary waves are generated, which travel into the new medium and contribute to the new wave-front.
- **Designing optical systems:** The Huygens-Fresnel principle is important for designing optical systems, such as lenses, mirrors, and telescopes. By understanding how light waves propagate through these systems, researchers can optimize their design to achieve the desired performance characteristics.
- **Understanding interference patterns:** The Huygens-Fresnel principle is used to understand interference patterns, which occur when two or more light waves interfere with each other. By understanding how the wavefronts of light propagate, researchers can predict the interference patterns that result and use them to study the properties of light.
- **Developing imaging techniques:** The Huygens-Fresnel principle is important for developing imaging techniques, such as holography and microscopy. By understanding how light waves propagate and interact with objects, researchers can develop imaging techniques that are able to visualize structures and materials at the microscopic and nanoscopic scale.

79. IDEAL GAS LAW

The Ideal Gas Law is a fundamental principle in thermodynamics that describes the behavior of an ideal gas. It relates the pressure (P), volume (V), absolute temperature (T), and number of molecules of a gas (n) to each other.

Mathematically, the Ideal Gas Law can be expressed as

$$PV = nRT$$

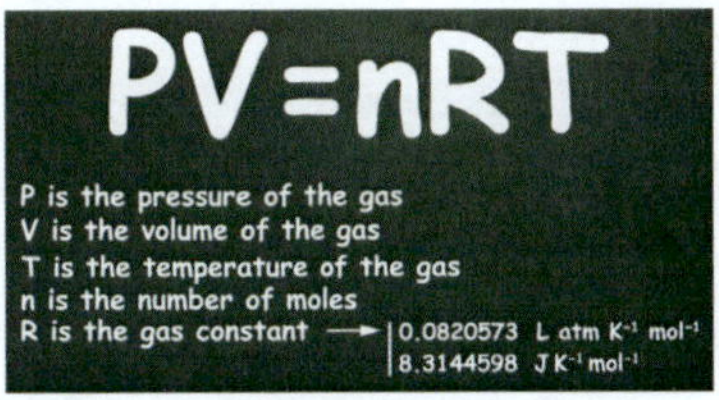

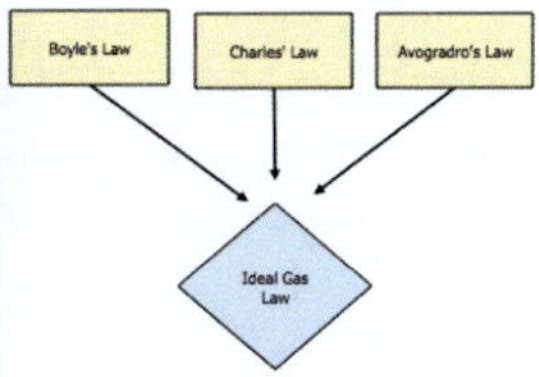

Applications of Ideal Gas Law

The Ideal Gas Law provides a powerful tool for understanding the behavior of ideal gases and has enabled the development of many important technologies.

- **Gas laws:** The Ideal Gas Law is the basis of many gas laws, such as Boyle's Law (which describes the relationship between the volume and pressure of a gas), Charles' Law (which describes the relationship between the volume and temperature of a gas), and Gay-Lussac's Law (which describes the relationship between the pressure and temperature of a gas).
- **Thermodynamics:** The Ideal Gas Law is used in thermodynamics to study the behavior of gases in various processes, such as isothermal, adiabatic, and isobaric processes.
- **Atmospheric science:** The Ideal Gas Law is used in atmospheric science to study the behavior of air and other gases in the Earth's atmosphere, including the relationship between pressure, temperature, and altitude.
- **Engineering:** The Ideal Gas Law is used in engineering to design and analyze gas-based systems, such as combustion engines, gas turbines, and air conditioning systems.

80. Inverse Square Law

The Inverse Square Law is a fundamental concept in physics that describes the decrease in intensity of a physical quantity, such as light or sound, as the distance from the source increases. It states that the intensity of a physical quantity decreases in proportion to the square of the distance from the source. The inverse square law states that the intensity of a physical quantity that emanates from a point source is inversely proportional to the square of the distance from the source. Mathematically, it is represented as:

$$I = k / d^2$$

where I is the intensity of the quantity, d is the distance from the source, and k is a constant that depends on the specific physical quantity being measured.

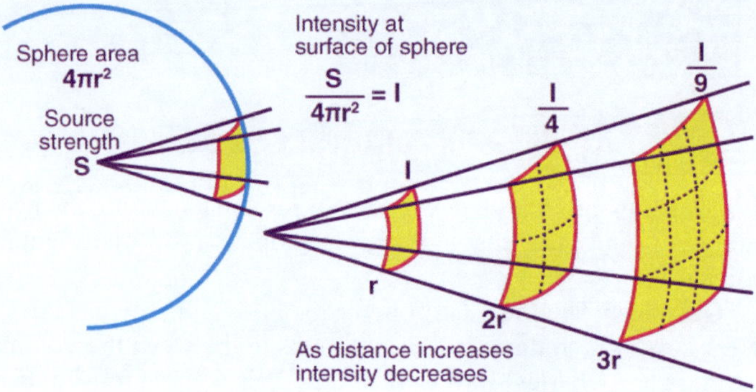

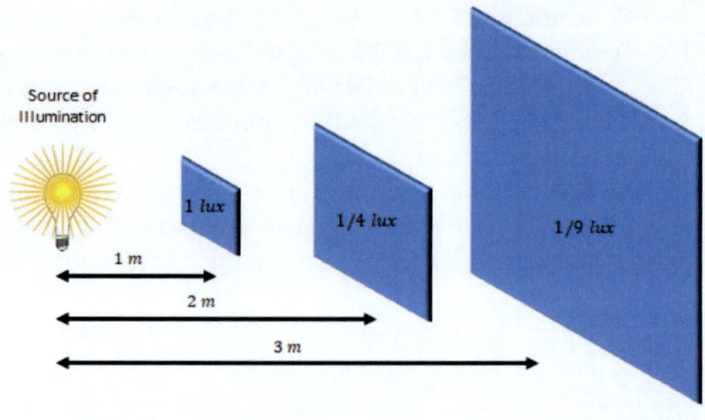

Applications of Inverse Square Law

Applications of the Inverse Square Law include:

- **Gravity:** The Inverse Square Law is applied in the study of gravity, which decreases in strength as the distance between two objects increases. This concept is used in the calculation of the gravitational force between objects, including planets, stars, and galaxies.
- **Light:** The Inverse Square Law is applied in the study of light, which decreases in intensity as the distance from the source increases. This concept is used in the calculation of the brightness of light sources, including stars, light bulbs, and flashlights.
- **Sound:** The Inverse Square Law is applied in the study of sound, which decreases in intensity as the distance from the source increases. This concept is used in the design of sound systems and in the calculation of the sound level at various distances from the source.
- **Radiation:** The Inverse Square Law is applied in the study of radiation, which decreases in intensity as the distance from the source increases. This concept is used in the design of radiation shielding, as well as in the measurement of radiation levels in various environments.
- **Electric fields:** The Inverse Square Law is applied in the study of electric fields, which decrease in strength as the distance from the source increases. This concept is used in the calculation of the strength of electric fields and in the design of electrical systems.

81. JOSEPH PROUST'S LAW OF DEFINITE COMPOSITION

Joseph Proust's law of definite composition, also known as the law of constant composition, states that a given compound always contains the same proportion of elements by mass. In other words, the ratio of the masses of the constituent elements in a compound is always the same, regardless of how the compound was prepared or where it came from. This law is a fundamental principle of chemistry and is used to understand the composition of many different types of compounds.

For example, water (H2O) always contains two hydrogen atoms and one oxygen atom, and the ratio of their masses is always the same. Similarly, carbon dioxide (CO2) always contains one carbon atom and two oxygen atoms, and the ratio of their masses is always the same. The law of definite composition is a cornerstone of modern chemistry and is essential for understanding the properties and behavior of chemical substances.

Joseph Proust
(Credit: https://wellcomeimages.org/indexplus/image/L0006972.html)

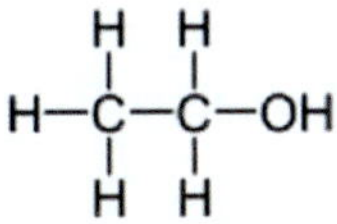

Ethyl Alcohol
C_2H_6O

2 atoms of C = 12 g/mol x 2 = 24
6 atoms of H = 1 g/mol x 6 = 6
<u>1 atom of O = 16 g/mol x 1 = 16</u>
Molecular weight of C_2H_6O = 46 g/mol

In 1 mole of C_2H_6O
% By Weight of C = 24/46 x 100% = 52%
% By Weight of H = 6/46 x 100% = 13%
% By Weight O = 16/46 x 100% = 35%

Ethyl alcohol will always have 52% carbon, 13 % Hydrogen and 35 % oxygen by weight

Applications of Joseph Proust's Law of Definite Composition

Joseph Proust's law of definite composition, also known as the law of constant proportion, states that a given compound always contains the same elements in the same proportion by weight. This law is fundamental to modern chemistry and has numerous practical applications, some of which are:

- **Chemical analysis:** Proust's law allows chemists to determine the composition of a compound. By analyzing the elements in a sample, they can calculate the exact proportions of each element in the compound.
- **Quality control:** Proust's law is used in the quality control of manufactured products. For example, the composition of a pharmaceutical drug must meet certain standards to ensure its safety and effectiveness.
- **Predicting reactions:** Proust's law can be used to predict the outcome of chemical reactions. If the proportion of reactants is known, the proportion of products can be calculated.
- **Identifying unknown compounds:** Proust's law is often used to identify unknown compounds. By analyzing the proportions of elements in a sample, chemists can identify the compound and determine its properties.
- **Industrial processes:** Proust's law is used in various industrial processes, such as the production of fertilizers, plastics, and chemicals. It allows manufacturers to precisely control the composition of their products.

82. JOULE'S LAW OF HEATING

James Prescott Joule, an English physicist, conducted experiments to establish the relationship between the flow of electric current through a conductor and the heat generated in the 1840s. He determined that the heat produced in a conductor is proportional to the square of the current passing through it, the resistance of the conductor, and the time for which the current flows. This relationship is expressed by the equation:

H = I^2 × R × t

where,
H represents the heat produced (measured in joules).
I is the current flowing through the conductor (measured in amperes).
R is the resistance of the conductor (measured in ohms).
t is the time for which the current flows (measured in seconds).

James Prescott Joule
(Credit: https://commons.wikimedia.org/wiki/File:James-Prescott-Joule.png)

Joule's Law (P=VI)

Power (P) (Measured in Watts W) is the rate of energy dissipation

Power is the number of Joules per Second converted from electrical energy to another form of energy.

$$P = VI \qquad \text{where } V = \frac{W}{Q} \qquad \text{and } I = \frac{Q}{t}$$

$$\text{therefore } P = \frac{W}{Q} \times \frac{Q}{t}$$

$$\text{giving } \quad P = \frac{W}{t}$$

Applications of Joule's Law of Heating

Joule's Laws have several practical applications, including:

- **Heating applications:** Joule's Laws are used in the design of heating elements for applications such as electric ovens, furnaces, and space heaters. By understanding the relationship between the current flowing through a conductor and the heat produced, engineers can design heating elements that are efficient and effective.
- **Electrical safety:** Joule's Laws are important in ensuring electrical safety. By understanding the relationship between the current flowing through a conductor and the heat produced, engineers can design circuits and electrical systems that are safe and do not overheat.
- **Energy conversion:** Joule's Laws are used in the design of electrical machines and motors that convert electrical energy into mechanical energy. By understanding the relationship between the electrical energy consumed and the heat produced, engineers can optimize the efficiency of these devices.
- **Material science:** Joule's Laws are used in material science to study the behavior of materials under different electrical and thermal conditions. By understanding the relationship between the current flowing through a material and the heat produced, scientists can study the properties of materials and how they respond to different stimuli.

83. JOULE-THOMSON EFFECT

The Joule-Thomson Effect is a thermodynamic phenomenon that describes the cooling or heating of a gas as it undergoes a throttling process. When a gas is expanded rapidly through a narrow orifice, it experiences a decrease in temperature due to the conversion of its internal energy into work. This effect is known as the Joule-Thomson Effect.

James Prescott Joule **William Thomson**
(Credit: https://wellcomeimages.org/indexplus/image/V0026628.html)

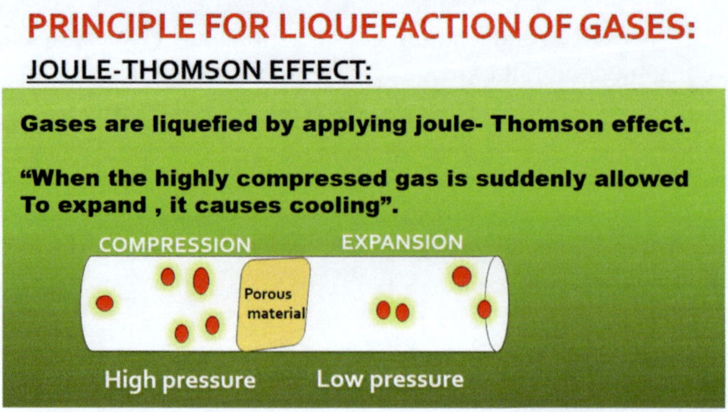

Applications of Joule-Thomson Effect

The Joule-Thomson Effect is a fundamental thermodynamic phenomenon that has many practical applications in various fields of science and engineering. It provides a powerful tool for producing low temperatures and removing impurities from gas streams, and has enabled the development of many important technologies, such as refrigeration systems and cryogenic applications. Some of the applications include:

- **Refrigeration:** The Joule-Thomson Effect is used in refrigeration systems to produce low temperatures. In a typical refrigeration cycle, a compressed gas is expanded through a valve or nozzle, causing it to cool down. The cooled gas is then used to cool a refrigerant, which is used to cool the surrounding environment.
- **Natural gas processing:** The Joule-Thomson Effect is used in natural gas processing to remove impurities from the gas. Natural gas typically contains impurities such as water vapor, carbon dioxide, and nitrogen. By using the Joule-Thomson Effect, these impurities can be removed from the gas stream, producing a cleaner and more usable gas product.
- **Cryogenics:** The Joule-Thomson Effect is used in cryogenic applications to produce low temperatures for scientific research and medical applications. Cryogenic temperatures are used in a variety of applications, such as cooling superconducting magnets, preserving biological samples, and studying the behavior of materials at low temperatures.
- **High-Pressure gas transmission:** The Joule-Thomson Effect is used in high-pressure gas transmission systems to prevent the buildup of ice or hydrates in pipelines. By expanding the gas through a Joule-Thomson valve, the temperature of the gas is reduced, preventing the formation of ice or hydrates that can clog the pipeline.

84. JURIN'S LAW

Jurin's law, named after the English mathematician and physicist James Jurin, relates to the capillary rise of a liquid in a narrow tube due to surface tension. According to Jurin's law, the height of the liquid column h in a capillary tube is inversely proportional to the diameter of the tube.

Mathematically, Jurin's law can be expressed as:

$$h = 2T \cos(\theta) / (\rho g D)$$

where T is the surface tension of the liquid, θ is the contact angle between the liquid and the tube, ρ is the density of the liquid, g is the acceleration due to gravity, and D is the diameter of the tube.

Jurin's law is a fundamental principle in capillary action and has many applications in various fields, such as in the design of microfluidic devices, inkjet printing, and paper chromatography.

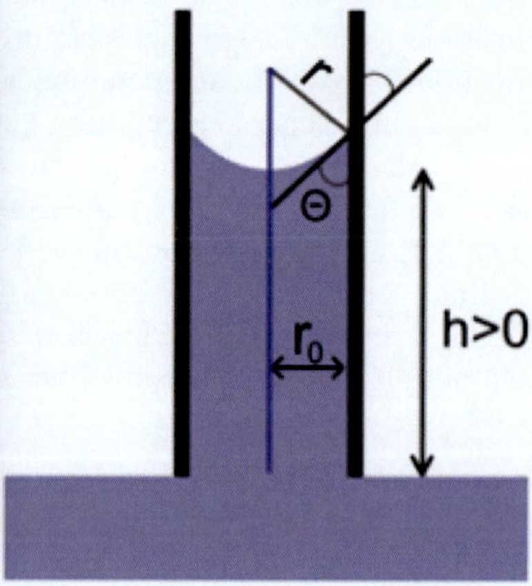

James Jurin
(Credit: http://trinitycollegechapel.com/about/memorials/shields/jurin/)

Applications of Jurin's Law

Jurin's law is a fundamental principle in the study of capillary action and fluid mechanics, and has many important applications in measuring surface tension, designing microfluidic devices, and understanding biological systems. It provides a framework for understanding the behavior of liquids in confined spaces and has helped to advance our understanding of fluid mechanics and its applications. Jurin's law, also known as the law of capillary rise, describes the relationship between the height of a liquid column and the surface tension and contact angle of the liquid. It has several important applications, including:

- **Measuring surface tension:** Jurin's law can be used to measure the surface tension of liquids. By measuring the height of a liquid column in a capillary tube and the contact angle of the liquid, researchers can calculate the surface tension of the liquid.
- **Understanding capillary action:** Jurin's law is important for understanding capillary action, which is the ability of liquids to rise up narrow tubes or through porous materials. By understanding the relationship between the height of a liquid column, the surface tension, and the contact angle, researchers can predict the behavior of liquids in capillary tubes and other confined spaces.
- **Designing microfluidic devices:** Jurin's law is important for designing microfluidic devices, which are small-scale systems that manipulate fluids for various applications. By understanding the behavior of liquids in narrow channels and confined spaces, researchers can design microfluidic devices that operate efficiently and reliably.
- **Understanding biological systems:** Jurin's law is used to study biological systems, such as the behavior of fluids in plant xylem vessels or the capillary action in the alveoli of the lungs. By understanding the relationship between the height of a liquid column, the surface tension, and the contact angle, researchers can gain insights into the behavior of fluids in biological systems.

85. KASHA'S RULE

Kasha's rule, named after the physical chemist Michael Kasha, states that the fluorescence emission of a molecule occurs from the lowest excited singlet state (S_1) to the ground state (S_0) of the molecule. In other words, the emission wavelength of a fluorescent molecule is independent of the excitation wavelength used to excite the molecule.

Kasha's rule is based on the fact that the fluorescence emission occurs on a very short timescale, typically within a few nanoseconds, which is much shorter than the timescale of intersystem crossing or internal conversion between excited states. Therefore, the fluorescence emission occurs from the lowest energy excited state, which is usually the S_1 state.

Kasha's rule has been used extensively in the field of fluorescence spectroscopy, where it provides a useful tool for the analysis and characterization of fluorescent molecules. It has also been used to develop a better understanding of the photophysical properties of molecules and their interactions with light.

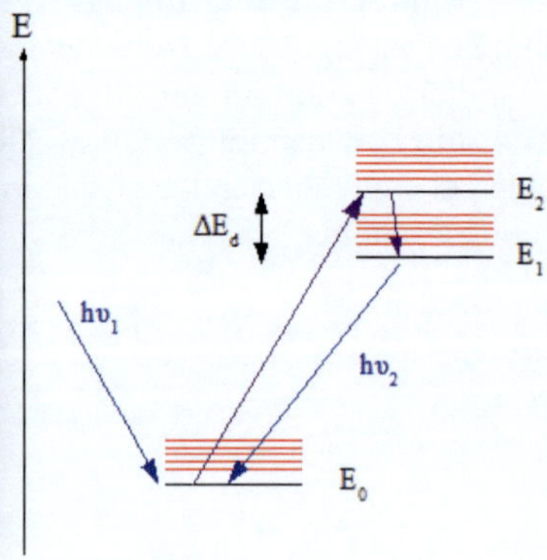

Michael Kasha
(Credit: https://www.gf.org/fellows/michael-kasha/)

Applications of Kasha's Rule

Kasha's rule, also known as the Kasha-Vavilov rule, describes the fluorescence emission maximum of a molecule as being independent of the excitation wavelength. Kasha's rule is a fundamental principle in the study of fluorescence and has many important applications in the design of fluorescent probes, imaging techniques, and the study of protein structure and function. It provides a framework for understanding the relationship between the excitation and emission wavelengths of fluorescent molecules and has helped to advance our understanding of the properties of fluorescence and its applications in biology and medicine. Some of the applications include:

- **Determining the emission maximum of fluorescent molecules:** Kasha's rule is used to predict the emission maximum of fluorescent molecules based on their excitation wavelength. By understanding the relationship between the excitation and emission wavelengths, researchers can design experiments to maximize the fluorescence signal and improve the sensitivity of fluorescence-based assays.
- **Characterizing the properties of fluorescent probes:** Kasha's rule is used to study the properties of fluorescent probes, such as their quantum yield, lifetime, and photostability. By understanding the fluorescence emission maximum of a probe and how it changes under different conditions, researchers can optimize the performance of the probe for various applications.
- **Developing fluorescence-based imaging techniques:** Kasha's rule is important for developing fluorescence-based imaging techniques, such as fluorescence microscopy and imaging flow cytometry. By understanding the properties of fluorescent probes and the relationship between the excitation and emission wavelengths, researchers can design imaging systems that produce high-quality images and accurate measurements.
- **Studying protein structure and function:** Kasha's rule is used to study the structure and function of proteins, particularly those that contain fluorescent probes as structural or functional probes. By understanding the fluorescence emission maximum of the probes and how it changes in response to changes in the protein structure or environment, researchers can gain insights into the behavior of proteins in various biological processes.

86. KELVIN PLANCK STATEMENT

The Kelvin-Planck statement is a principle of thermodynamics that describes the limitations of heat engines. The statement was formulated by Lord Kelvin and was later modified and restated by Max Planck.

The Kelvin-Planck statement can be expressed as follows:

"It is impossible for any device that operates on a cycle to receive heat from a single reservoir and produce a net amount of work."

In simpler terms, it means that it is impossible to convert all the heat supplied to a heat engine into work.

This statement implies that it is impossible to create a heat engine that can convert all the heat it absorbs into work without producing any waste heat or transferring heat to a cooler reservoir. This is because the **second law of thermodynamics** states that heat naturally flows from a hotter object to a cooler object.

In other words, the Kelvin-Planck statement implies that it is impossible to build a heat engine that can operate continuously while producing useful work without any input of energy. This is an important principle in thermodynamics and has significant implications for the design and efficiency of heat engines.

The Kelvin-Planck statement has important implications for the design and operation of heat engines, such as steam turbines and internal combustion engines. It also has wider implications for energy conversion processes and the study of the universe as a whole.

Lord Kelvin

Max Planck

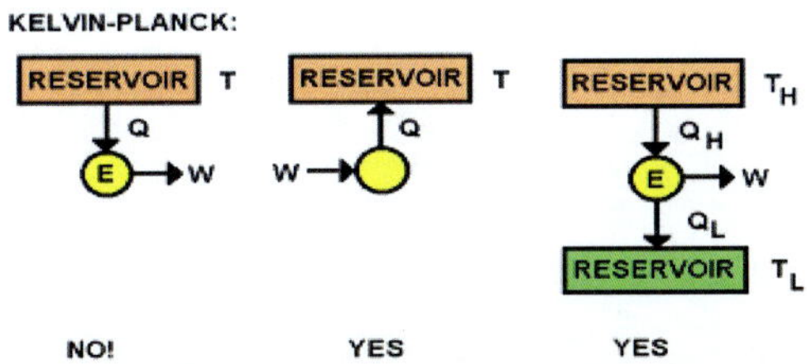

Applications of Kelvin-Planck Statement

The Kelvin-Planck statement has important implications for the design and operation of energy conversion systems, and it provides a fundamental framework for understanding the limitations of thermodynamic processes. The Kelvin-Planck statement has several important applications in the field of thermodynamics and engineering. Some of these applications are:

- **Limitations of heat engines:** The Kelvin-Planck statement states that it is impossible for a heat engine to convert all the heat it receives into work. This statement sets a fundamental limit on the efficiency of heat engines. Engineers use this principle to design and optimize heat engines, such as internal combustion engines, steam turbines, and refrigeration systems.
- **Design of refrigeration systems:** The Kelvin-Planck statement also has implications for the design and operation of refrigeration systems. Refrigeration systems use a refrigerant to transfer heat from a cold environment to a hot environment. The efficiency of these systems is limited by the Kelvin-Planck statement, which requires that some heat must be rejected to a hotter environment.
- **Evaluation of energy systems:** The Kelvin-Planck statement provides a useful framework for evaluating the efficiency of energy systems. By comparing the energy output of a system to the energy input, engineers can calculate the efficiency of the system and determine whether it is operating within the limitations of the Kelvin-Planck statement.
- **Thermodynamic cycles:** The Kelvin-Planck statement is closely related to the concept of thermodynamic cycles. A thermodynamic cycle is a sequence of processes that return a system to its initial state. The Kelvin-Planck statement sets a fundamental limit on the efficiency of any thermodynamic cycle, which is known as the Carnot cycle. Engineers can use the principles of the Carnot cycle to optimize the efficiency of heat engines and other energy conversion processes.

87. KEPLER'S LAWS OF PLANETARY MOTION

Kepler's laws of planetary motion are a set of three laws discovered by the German astronomer Johannes Kepler in the early 17th century. These laws describe the motion of planets around the sun and are considered to be among the most important laws in the history of astronomy. The three laws are:

- **The law of ellipses:** Each planet orbits the sun in an ellipse with the sun at one focus of the ellipse.
- **The law of equal areas:** The line connecting a planet to the sun sweeps out equal areas in equal times as the planet moves around the ellipse. This means that a planet moves faster when it is closer to the sun and slower when it is farther away.
- **The law of harmonies:** The square of the orbital period of a planet is proportional to the cube of the semi-major axis of its orbit. In simpler terms, the farther a planet is from the sun, the longer its orbital period.

Kepler's laws were a major milestone in the development of modern astronomy and played a key role in the development of Newton's laws of motion and the law of universal gravitation. They continue to be important in the study of the solar system and have been used to discover numerous exoplanets orbiting other stars.

Johannes Kepler

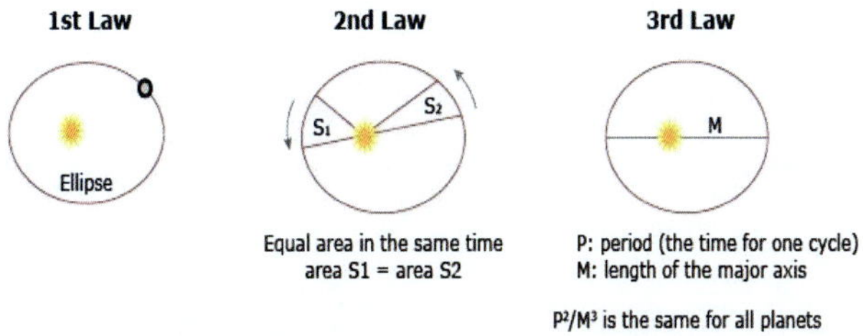

Applications of Kepler's Laws of Planetary Motion

Kepler's laws of planetary motion have numerous applications in astronomy, space exploration, and other fields. They provide a mathematical framework for understanding the motion of celestial objects and have helped to advance our understanding of the universe and our place in it. Some of the most notable applications include:

- **Predicting the motion of planets and other celestial objects:** Kepler's laws allow astronomers to accurately predict the motion of planets and other celestial objects. By understanding how planets move around the sun, astronomers can make predictions about the timing and location of astronomical events such as eclipses and planetary transits.
- **Finding exoplanets:** Kepler's laws are used to discover and study exoplanets, which are planets outside our solar system. By observing the motion of a star over time, astronomers can detect the presence of exoplanets and determine their size, orbit, and other properties.
- **Navigation:** Kepler's laws have been used in the development of navigation systems, such as the Global Positioning System (GPS). The laws provide a mathematical framework for determining the position and velocity of satellites in orbit around the Earth, which is essential for accurate navigation and communication.
- **Orbital mechanics:** Kepler's laws are a key part of orbital mechanics, which is the study of how objects move in orbit around each other. Understanding the laws of motion and gravity that govern the motion of planets, moons, and other celestial objects is essential for space exploration and the design of spacecraft.
- **Planetary science:** Kepler's laws are used in the study of planetary science, including the formation and evolution of planets and the search for life beyond Earth. By understanding the motion and orbits of planets, scientists can gain insights into the physical and chemical processes that shape the solar system and other planetary systems.

88. KIRCHHOFF'S LAW

Kirchhoff's laws are a set of fundamental laws in electrical circuit analysis that govern the behavior of electrical circuits. Gustav Kirchhoff, a German physicist, first introduced these laws in 1845. There are two Kirchhoff's laws:

- **Kirchhoff's current law (KCL):** Kirchhoff's first Law, also known as the Kirchhoff's Current Law (KCL), states that the total current flowing into a junction in an electrical circuit is equal to the total current flowing out of the junction. In other words, the algebraic sum of the currents at any junction in an electrical circuit is zero. This law is based on the principle of conservation of charge.
- **Kirchhoff's voltage law (KVL):** Kirchhoff's second law, also known as Kirchhoff's voltage law (KVL), states that the algebraic sum of all the voltage drops around any closed loop in a circuit must be zero. This law is based on the principle of conservation of energy, which states that the total energy in a closed system remains constant.

In simpler terms, KVL means that the total voltage supplied in a closed loop circuit must be equal to the total voltage dropped across all the resistive elements in the loop.

Gustav Kirchhoff

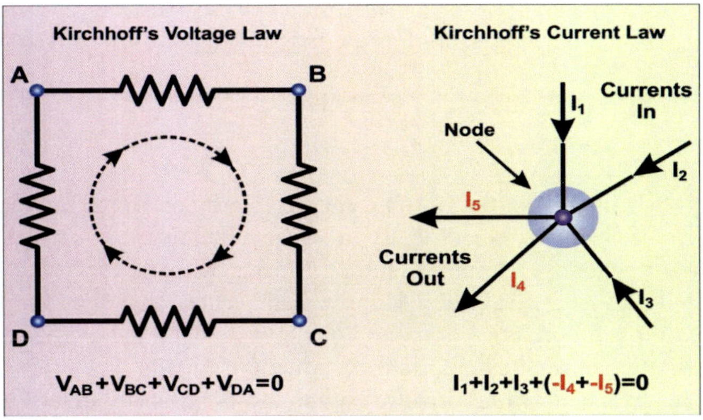

Applications of Kirchhoff's Law

Kirchhoff's laws are fundamental laws in electrical circuit analysis that have a wide range of applications in engineering and physics. They are essential in the design and optimization of complex electrical systems and are a cornerstone of modern electrical engineering. They allow engineers and physicists to calculate the currents and voltages in a circuit, as well as to predict the behavior of the circuit under different conditions. Some applications of Kirchhoff's laws include:

- **Circuit analysis:** Kirchhoff's laws are used to analyze complex electrical circuits and to calculate the currents and voltages at various points in the circuit. This information is essential in the design and optimization of electrical circuits.
- **Sensor design:** Kirchhoff's laws are used in the design of sensors that measure electrical quantities such as current and voltage. Sensors based on Kirchhoff's laws are used in a wide range of applications, from industrial control systems to medical devices.
- **Power distribution:** Kirchhoff's laws are used in the design of power distribution systems, such as electrical power grids. By analyzing the currents and voltages in the system, engineers can ensure that the system is operating efficiently and safely.
- **Electronic design:** Kirchhoff's laws are used in the design of electronic circuits, such as amplifiers and filters. By understanding the behavior of circuits under different conditions, engineers can optimize their designs for specific applications.

89. KOPP'S RULE

Kopp's rule, named after the German chemist Hermann Kopp, is a principle in the field of chemistry, specifically in organic chemistry. It is named after the German chemist Hermann Franz Moritz Kopp (1817-1892) who formulated this empirical rule. Kopp's Rule states that the molar heat capacity of a solid organic compound can be estimated by adding together the contributions from its individual constituent elements. The rule assumes that the heat capacity of an organic compound is primarily determined by the masses and types of atoms present in its chemical structure.

According to Kopp's Rule, the molar heat capacity of an organic compound (C) can be approximated as:

$$C = \sum_{i=1}^{N} C_i f_i,$$

where, N is the total number of compound constituents, and Ci and fi denote the specific heat and mass fraction of the i-th constituent.

It is important to note that Kopp's Rule is an empirical rule and has limitations. It provides only an estimation of the molar heat capacity and may not be accurate for all organic compounds. Additionally, it assumes that the heat capacity contributions of the constituent elements are independent of each other, which may not always hold true.

Kopp's law was first proposed in 1842 and was an early attempt to establish a relationship between the physical properties of elements and their atomic weights. While Kopp's law was later found to have some limitations and exceptions, it remains an important early contribution to the field of thermodynamics and the study of the physical properties of matter.

Today, we know that the specific heat of a solid depends not only on its atomic weight but also on other factors such as the crystal structure, electronic structure, and bonding in the solid.

$$(C_p)_{Ca(OH)2} = (C_{pa})_{Ca} + 2\,(C_{pa})_{O} + 2\,(C_{pa})_{H}$$

$$= 26 + (2x17) + (2x9.6) = 79\ J/mol.^{\circ}C$$

Hermann Kopp
(Credit:https://commons.wikimedia.org/wiki/File:Hermann_Franz_Moriz_Kopp_(HeidICON_36744).jpg)

Applications of Kopp's Law

Kopp's law, also known as the law of heat summation or the law of specific heats, has several important applications in the fields of thermodynamics, physical chemistry, and materials science. By understanding the relationship between molecular weight and specific heat, scientists can make predictions about the thermal behavior of substances and design materials and processes with specific thermal properties. Some notable applications include:

- **Determination of specific heat:** Kopp's law is used to determine the specific heat of various substances, which is a crucial property for understanding their thermal behavior. By measuring the molecular weight of a substance, Kopp's law can be used to predict its specific heat.
- **Prediction of thermodynamic properties:** Kopp's law can be used to predict the thermodynamic properties of substances, such as their enthalpy and entropy, based on their molecular weight. This allows scientists to model and predict the behavior of substances under different conditions, such as temperature and pressure.
- **Development of materials science:** Kopp's law has played an important role in the development of materials science, as it allows scientists to predict the thermal behavior of various materials. This is useful in designing materials with specific thermal properties for applications such as insulation and heat dissipation.
- **Understanding combustion and energy production:** Kopp's law has important applications in the field of combustion, as it can be used to predict the heat of combustion for various substances. This is useful in understanding the energy production potential of different fuels, and in optimizing combustion processes for efficiency and safety.

90. LAGRANGIAN POINT

Lagrangian points, also known as Lagrange points, are five points in space where the gravitational forces of two large bodies, such as a planet and a moon or a planet and a star, balance out the centrifugal force of a smaller object, such as a spacecraft or satellite, allowing it to remain in a stable position relative to the two larger bodies.

The five Lagrangian points are labeled L1, L2, L3, L4, and L5. L1 is located between the two larger bodies, while L2 and L3 are located on the opposite side of the smaller object. L4 and L5 are located in the orbit of the smaller object, forming equilateral triangles with the two larger bodies.

Joseph-Louis Lagrange

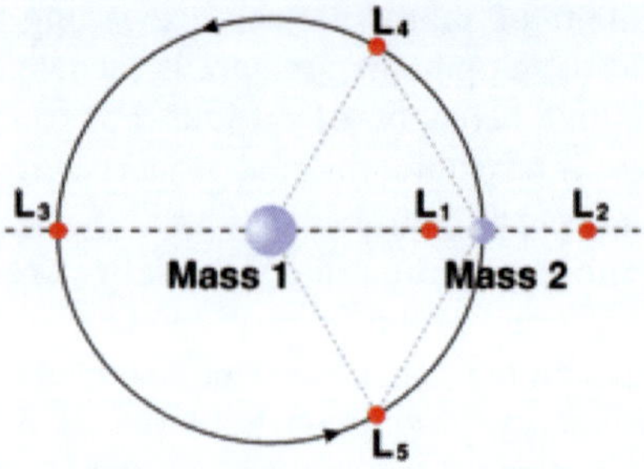

Applications of Lagrangian Point

Lagrangian points offer a unique opportunity for space exploration and astronomy, providing a stable platform for observations and serving as a strategic point for interplanetary travel. The applications include:

- **Satellite positioning:** Satellites can be placed at Lagrangian points to provide a stable platform for astronomical observations or to monitor the Earth's environment. This is useful for monitoring the Earth's environment, observing distant objects in the universe, and conducting experiments in microgravity. For example, the Solar and Heliospheric Observatory (SOHO) satellite is located at L1, where it continuously monitors the Sun's activity.
- **Space telescopes:** Lagrangian points provide a stable location for space telescopes, allowing them to maintain a fixed position relative to the Earth and the Sun. This is important for conducting long-term observations of distant objects in the universe, such as galaxies, stars, and exoplanets. The Hubble Space Telescope, for example, orbits at L2.
- **Interplanetary travel:** Lagrangian points can be used as a transfer point for spacecraft traveling between different planets. By using the gravitational forces at these points, spacecraft can save fuel and travel more efficiently. This is particularly useful for missions that require multiple gravitational assists, such as the Cassini mission to Saturn.
- **Asteroid exploration:** Lagrangian points are also potential targets for asteroid exploration missions. Asteroids can be placed in stable orbits around Lagrangian points, making them easier to study and potentially mine for resources.
- **Space weather monitoring:** Satellites placed at Lagrangian points, such as the Solar and Heliospheric Observatory (SOHO) at L1, can monitor the Sun's activity and provide early warning of potentially dangerous solar storms.
- **Fundamental physics:** Lagrangian points are also used in experiments to test fundamental physics theories, such as the equivalence principle of general relativity.

91. LAMBERT'S COSINE LAW

Lambert's cosine law, also known as Lambert's law or Lambert's cosine emission law, is a fundamental principle in optics that describes the relationship between the intensity of light emitted by a surface and the angle of incidence of the light on that surface.

The law states that the radiant intensity or brightness of a surface varies with the cosine of the angle between the direction of the emitted radiation and the normal to the surface. In other words, the intensity of light emitted by a surface is proportional to the cosine of the angle between the direction of the emitted light and the surface normal.

Mathematically, the rule can be expressed as:

$$I = I_0 \cos\theta$$

where I is the intensity of light, I_0 is the intensity of incident light, and θ is the angle between the direction of incident light and the normal to the surface.

This law is important in a number of fields, including radiometry, photography, lighting design, computer graphics, and solar energy. It is particularly useful in understanding the behavior of light in 3D environments and for optimizing the placement and orientation of lighting sources or solar panels to maximize their effectiveness.

Johann Heinrich Lambert

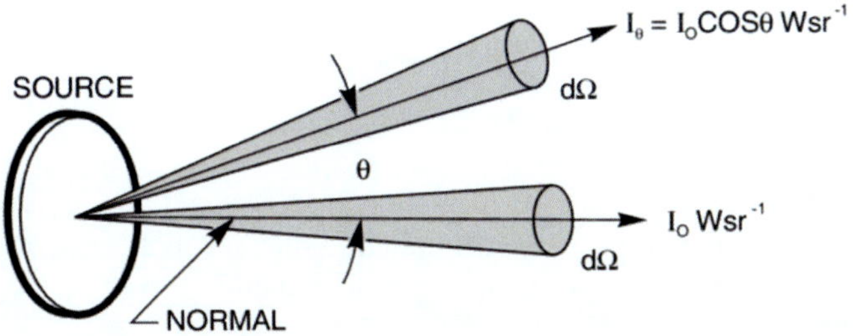

Applications of Lambert's Cosine Law

Lambert's cosine law has numerous applications in fields such as radiometry, photography, lighting design, computer graphics, and solar energy, among others. This law is widely used in various fields, including:

- **Radiometry:** Lambert's cosine law is fundamental to radiometry, which is the measurement of radiation. It is used to calculate the amount of radiation received by a detector from a source, taking into account the angle of incidence and the reflectivity of the surface.
- **Photography:** In photography, Lambert's cosine law is used to calculate the amount of light that falls on the film or sensor, depending on the angle of incidence and the surface reflectivity. This helps to determine the correct exposure settings for a particular shot.
- **Lighting design:** Lambert's cosine law is used in lighting design to determine the amount of light emitted by a light source and the angle at which the light is emitted. This helps to design lighting systems that provide optimal illumination for a particular space.
- **Computer graphics:** In computer graphics, Lambert's cosine law is used to simulate the behavior of light in a 3D environment. This helps to create realistic lighting effects in video games, movies, and other digital media.
- **Solar energy:** Lambert's cosine law is used to calculate the amount of solar radiation received by a solar panel, taking into account the angle of incidence and the reflectivity of the panel. This helps to optimize the placement and orientation of solar panels to maximize energy output.

92. LAMM EQUATION

The Lamm equation is a mathematical equation that describes the movement of particles in a fluid during centrifugation. It was developed by Ole Albert Lamm.

The Lamm equation is a partial differential equation that relates the radial and axial movement of a particle in a centrifugal field to its sedimentation coefficient, which is a measure of the rate at which the particle moves in the centrifugal field. The equation takes into account the effects of both sedimentation and diffusion on the movement of the particle.

$$\frac{\partial c}{\partial t} = D\left[\frac{\partial^2 c}{\partial r^2} + \frac{1}{r}\frac{\partial c}{\partial r}\right] - s\omega^2\left[r\frac{\partial c}{\partial r} + 2c\right]$$

where c is the solute concentration, t and r are the time and radius, and the parameters D, s, and ω represent the solute diffusion constant, sedimentation coefficient and the rotor angular velocity, respectively.

The Lamm equation has many applications in biochemistry, where it is used to study the behavior of macromolecules such as proteins and nucleic acids in solution. By measuring the sedimentation coefficient of a macromolecule in a centrifugal field and applying the Lamm equation, researchers can obtain information about its size, shape, and molecular weight. The Lamm equation has also been used in the study of colloidal systems.

Ole Albert Lamm

(Credit: https://sok.riksarkivet.se/sbl/Presentation.aspx?id=10964)

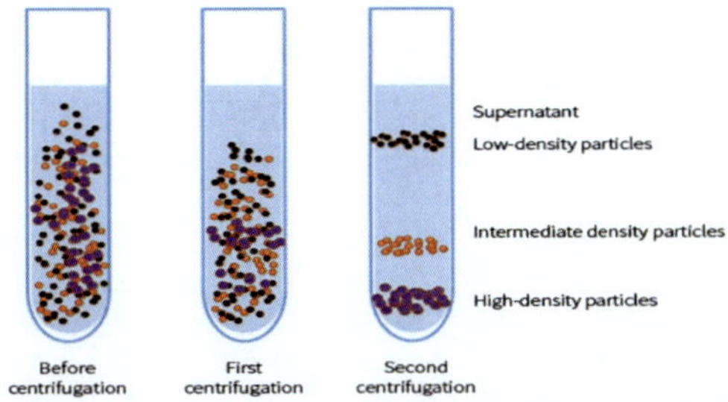

Applications of Lamm Equation

The Lamm equation is a mathematical model that describes the movement of molecules in a centrifugal field. Lamm equation is a powerful tool for analyzing the sedimentation behavior of macromolecules and has numerous applications in biophysics, drug discovery, and other areas of research. Its versatility and accuracy make it a valuable tool for understanding the structure, function, and interactions of complex biomolecules.

- **Ultracentrifugation:** The Lamm equation is widely used in ultracentrifugation experiments to analyze the sedimentation behavior of macromolecules, such as proteins, DNA, and RNA. By using the Lamm equation, researchers can calculate the molecular weight, shape, and size distribution of these molecules.
- **Density gradient centrifugation:** The Lamm equation is also used in density gradient centrifugation experiments, which separate molecules based on their buoyant density. By using the Lamm equation, researchers can accurately predict the sedimentation behavior of molecules in different density gradients, allowing for precise separation and analysis.
- **Drug discovery:** The Lamm equation has applications in drug discovery and development, as it can be used to analyze the binding properties of drugs and their interactions with macromolecules. By using ultracentrifugation experiments and the Lamm equation, researchers can determine the binding affinity, stoichiometry, and kinetics of drug-macromolecule interactions.
- **Biophysics:** The Lamm equation has numerous applications in biophysics, as it can be used to study the structure and dynamics of macromolecules. For example, the Lamm equation can be used to determine the shape and size of protein complexes and to analyze the conformational changes of proteins in response to ligand binding or other stimuli.

93. LANGMUIR EQUATION

The Langmuir equation, named after Irving Langmuir, is a model that describes the adsorption of gas molecules onto a solid surface. It is based on the assumption that the adsorption occurs through a reversible process, in which gas molecules in the surrounding environment adhere to the surface of the solid. The Langmuir equation can be written as:

$$\theta = K\,p / (1 + K\,p)$$

where θ is the fractional coverage of the surface by adsorbed gas molecules, p is the partial pressure of the gas in the surrounding environment, and K is the Langmuir constant, which is a measure of the strength of the interaction between the gas molecules and the solid surface.

The Langmuir equation is often used to model the behavior of adsorbents, such as activated carbon and zeolites, which are commonly used in gas separation and purification processes. By fitting experimental data to the Langmuir equation, researchers can obtain information about the adsorption capacity and selectivity of different adsorbents, which is important for optimizing their use in industrial applications.

The Langmuir equation has also been applied in the study of surface chemistry and catalysis, where it is used to model the adsorption of reactant molecules onto the surface of a catalyst. In this context, the Langmuir constant is related to the equilibrium constant of the adsorption reaction, and can be used to predict the rate of the catalytic reaction.

Irving Langmuir

(Credit: https://commons.wikimedia.org/wiki/File:IrvingLangmuir.jpg)

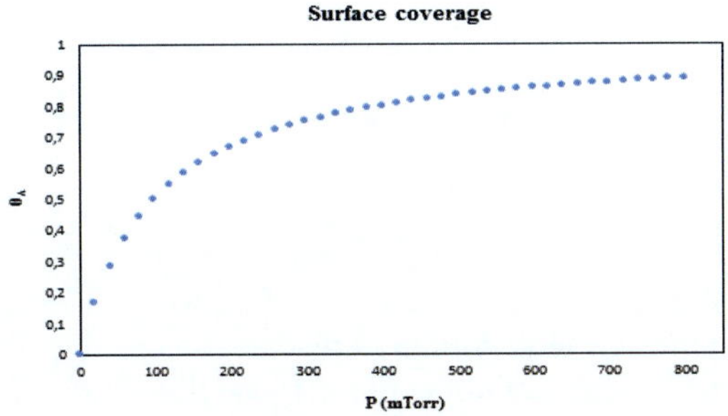

Applications of Langmuir Equation

The Langmuir equation is a mathematical model that describes the adsorption of molecules onto a solid surface. It has many applications in the fields of surface chemistry, materials science, and chemical engineering. Here are some notable applications of the Langmuir equation:

- **Adsorption studies:** The Langmuir equation is commonly used to model the adsorption of molecules onto surfaces, such as the adsorption of gases onto solid catalysts in industrial processes. By using the Langmuir equation, researchers can determine the maximum amount of molecules that can be adsorbed onto a surface and the affinity of the molecules for the surface.
- **Surface characterization:** The Langmuir equation can also be used to study the properties of surfaces, such as their porosity and surface area. By measuring the adsorption of molecules onto a surface and fitting the data to the Langmuir equation, researchers can determine the surface area and pore size distribution of the material.
- **Biosensors:** The Langmuir equation is used in the design of biosensors, which are devices that detect the presence of biomolecules, such as proteins or DNA. By immobilizing biomolecules onto a solid surface and measuring their interaction with target molecules, researchers can use the Langmuir equation to calculate the binding affinity and kinetics of the interaction.
- **Water treatment:** The Langmuir equation has applications in the field of water treatment, where it is used to model the adsorption of contaminants, such as heavy metals, onto surfaces. By using the Langmuir equation, researchers can determine the maximum adsorption capacity of a material for a particular contaminant, which is useful in designing effective water treatment processes.

94. LAPLACE TRANSFORM

The Laplace transform is a mathematical technique used to transform a function of time into a function of complex frequency. It is named after the French mathematician Pierre-Simon Laplace.

The Laplace transform of a function f(t) is defined as:

$$F(s) = \int_{0}^{+\infty} f(t) \cdot e^{-s \cdot t} \cdot dt$$

where s is a complex frequency parameter, and the integral is taken over the range of t from zero to infinity. The function F(s) is called the Laplace transform of f(t).

The Laplace transform is a powerful tool in applied mathematics and engineering, as it allows for the analysis of systems that are described by differential equations. By taking the Laplace transform of a differential equation, it is possible to convert the equation into an algebraic equation, which is often easier to solve.

The Laplace transform is also useful in signal processing and control theory, where it is used to analyze and design systems that operate in the frequency domain. It is a fundamental tool in the analysis and design of control, communication and signal processing systems.

Pierre-Simon Laplace
(Credit: Portrait by Johann Ernst Heinsius;
https://www.pinterest.ru/pin/167759154858564779/)

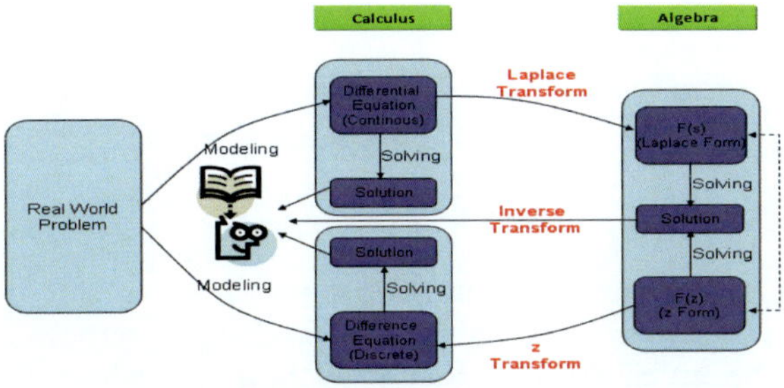

Applications of Laplace Transform

The Laplace transform is a mathematical technique that has a wide range of applications in various fields, including engineering, physics, economics, and finance. Some notable applications are:

- **Control systems:** The Laplace transform is widely used in the design and analysis of control systems. By converting differential equations into algebraic equations, Laplace transform simplifies the process of solving complex control problems.
- **Circuit analysis:** The Laplace transform is an essential tool for analyzing electrical circuits. It is used to determine the steady-state behavior of circuits and to calculate the transfer functions of filters and amplifiers.
- **Signal processing:** The Laplace transform is used extensively in signal processing. It is used to transform signals between the time domain and frequency domain, allowing engineers to analyze signals in different domains.
- **Differential equations:** The Laplace transform can be used to solve differential equations, including ordinary and partial differential equations. This technique provides an alternative method to solve these equations, especially for those equations that are difficult to solve using traditional methods.
- **Probability and statistics:** The Laplace transform is used in probability and statistics to analyze the distribution of random variables. It provides a way to transform the probability density function of a random variable into a moment-generating function.
- **Finance:** The Laplace transform has applications in finance for option pricing and risk management. It is used to calculate the expected value of future cash flows and to determine the risk associated with different investment options.

95. Larmor Formula

The Larmor formula, named after the Irish physicist Joseph Larmor, gives the rate at which energy is radiated by a charged particle undergoing acceleration. Specifically, the formula gives the power radiated per unit time by an accelerating charged particle, such as an electron, in terms of its charge, acceleration, and velocity.

The Larmor formula is given by:

$$P = (2/3) \times (q^2 \times a^2 / c^3)$$

where P is the power radiated per unit time, q is the charge of the particle, a is its acceleration, and c is the speed of light in vacuum.

The Larmor formula is an important result in classical electromagnetism and is used to describe a wide range of phenomena, including the emission of electromagnetic radiation by accelerating charged particles in atoms and molecules, as well as in high-energy astrophysical environments such as pulsars and black holes.

Joseph Larmor

$$P = \frac{2}{3}\frac{q^2 a^2}{4\pi\varepsilon_0 c^3} = \frac{q^2 a^2}{6\pi\varepsilon_0 c^3}$$

$$P = \frac{2}{3}\frac{q^2 a^2}{c^3} \text{ (cgs units)}$$

Applications of Larmor Formula

Larmor formula, also known as Larmor's theorem, is a fundamental principle in electromagnetism that describes the rate at which an accelerated charged particle radiates energy in the form of electromagnetic waves. It has numerous applications in physics, astrophysics, and engineering. Here are some notable applications:

- **Particle accelerators:** Larmor formula is used to calculate the power radiated by accelerated charged particles in particle accelerators, such as synchrotrons and cyclotrons. This is important in designing and optimizing the performance of these devices.
- **Astrophysics:** Larmor formula is used to understand the behavior of charged particles in astrophysical environments, such as the magnetic fields around stars and planets. It is also used to calculate the energy loss and lifetime of cosmic ray particles as they travel through interstellar space.
- **Radiation therapy:** Larmor formula is used in medical radiation therapy to calculate the dose of radiation absorbed by tissues in the body. This is important in designing treatment plans for cancer patients.
- **Electrical engineering:** Larmor formula is used in the design of electrical machines, such as generators and transformers, to calculate the amount of energy lost due to electromagnetic radiation.
- **Radio broadcasting:** Larmor formula is used to calculate the power radiated by radio antennas, and to design and optimize radio broadcasting systems.

96. LAW OF CONSERVATION OF ENERGY

The law of conservation of energy is a fundamental law of physics that states that the total amount of energy in a closed system remains constant over time. In other words, energy can neither be created nor destroyed; it can only be transformed from one form to another.

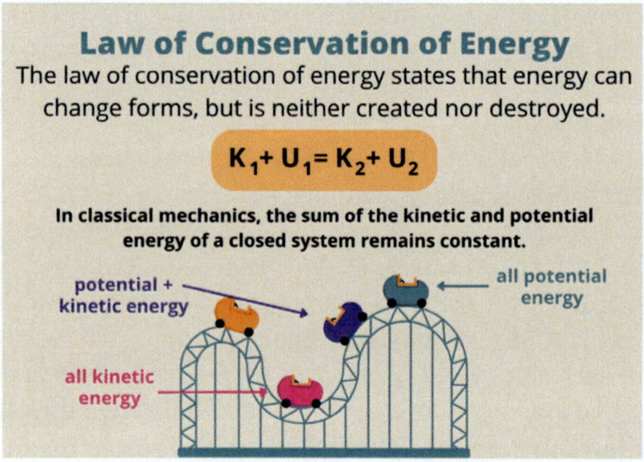

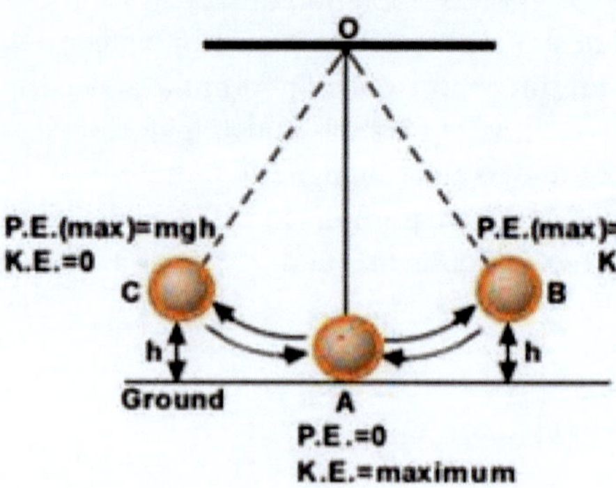

Applications of Law of Conservation of Energy

The law of conservation of energy is a fundamental principle in physics that has many practical applications in our daily lives, as well as in numerous scientific fields. Some of the most common applications of the law of conservation of energy include:

- **Mechanical systems:** The law of conservation of energy is used to analyze the motion of objects in mechanical systems. By knowing the initial energy of an object and the work done on it, it is possible to calculate its final energy and velocity.
- **Thermodynamics:** The law of conservation of energy is a key principle in thermodynamics, which is the study of energy and heat transfer. In particular, it is used to analyze the flow of heat between objects and the changes in temperature that result.
- **Electricity and magnetism:** The law of conservation of energy is used to analyze the behavior of electric and magnetic fields. By knowing the initial energy of a field and the work done on it, it is possible to calculate its final energy and the resulting electromagnetic effects.
- **Nuclear physics:** The law of conservation of energy is a fundamental principle in nuclear physics. It is used to calculate the energy released in nuclear reactions, and to analyze the behavior of radioactive decay.
- **Renewable energy:** The law of conservation of energy is central to the development and use of renewable energy sources such as solar, wind, and hydro power. These sources of energy are all based on the conversion of energy from one form to another, while maintaining the total amount of energy.
- **Energy efficiency:** The law of conservation of energy is used to design and improve energy-efficient devices and systems, such as buildings, appliances, and vehicles. By minimizing energy losses and optimizing energy use, these systems can save energy and reduce costs.
- **Chemical reactions:** The law of conservation of energy is used to understand and predict the behavior of chemical reactions, including the release or absorption of energy during these reactions.
- **Astronomy:** The law of conservation of energy is used to study the behavior of stars and galaxies. By analyzing the energy released by these objects and the effects of gravity, astronomers can better understand the origins and evolution of the universe.
- **Environmental science:** The law of conservation of energy is used to understand and address environmental issues such as climate change and pollution. By analyzing energy flows and using energy-efficient technologies, it is possible to reduce greenhouse gas emissions and protect the environment.

97. Law of Conservation of Linear Momentum

The Law of Conservation of Linear Momentum is a fundamental principle in physics that states that the total momentum of a system of objects remains constant if no external forces act on the system. In other words, if the net external force on a system of objects is zero, then the total momentum of the system remains constant.

Momentum is a measure of an object's motion, and is defined as the product of its mass and velocity. When two or more objects interact with each other, they can transfer momentum to each other. According to the Law of Conservation of Linear Momentum, the total momentum of the system before the interaction is equal to the total momentum after the interaction, as long as no external forces act on the system.

Applications of Law of Conservation of Linear Momentum

The Law of Conservation of Linear Momentum has many practical applications in physics and engineering. For example, it can be used to analyze the motion of objects in collisions, both elastic and inelastic, and to calculate the velocity and direction of the objects after the collision. It is also used in the design and analysis of rockets, satellites, and other spacecraft, where it is important to conserve momentum in order to control the motion of the spacecraft. Some of the applications of this law are:

- **Projectile motion:** The law of conservation of linear momentum is used in the analysis of projectile motion, such as the motion of a ball being thrown or a projectile being fired from a cannon. By understanding the relationship between the momentum of the projectile and the forces acting on it, engineers can design more accurate and efficient projectile systems.
- **Collision analysis:** The law of conservation of linear momentum is used in the analysis of collisions, such as car accidents. By analyzing the momentum of the vehicles before and after the collision, investigators can determine the cause of the accident and calculate the forces involved.
- **Rockets and spacecraft:** The law of conservation of linear momentum is used in the design of rockets and spacecraft, which rely on the conservation of momentum to propel themselves through space. By expelling mass in one direction, the rocket gains momentum in the opposite direction, allowing it to move forward.
- **Fluid dynamics:** The law of conservation of linear momentum is used in the analysis of fluid flow, such as the flow of water through a pipe. By understanding the momentum of the fluid and the forces acting on it, engineers can design more efficient and reliable fluid systems.

98. LAW OF CONSERVATION OF MASS

The Law of Conservation of Mass is a fundamental principle in chemistry and physics that states that mass is neither created nor destroyed in any chemical or physical process. This means that the total mass of the reactants in a chemical reaction is equal to the total mass of the products that are formed.

The Law of Conservation of Mass is also known as the Law of Mass Conservation or the Principle of Mass Conservation.

This law was first proposed by French chemist Antoine Lavoisier in the late 18th century, who conducted experiments that showed that the total mass of substances in a chemical reaction remains constant, even if the substances are transformed into different states, such as solids, liquids, or gases.

The Law of Conservation of Mass is a cornerstone of modern chemistry and physics, and it is used in a wide range of applications. In summary, the Law of Conservation of Mass is a fundamental principle that describes the relationship between matter and energy, and it is an essential concept in many areas of science and engineering.

Antoine Lavoisier

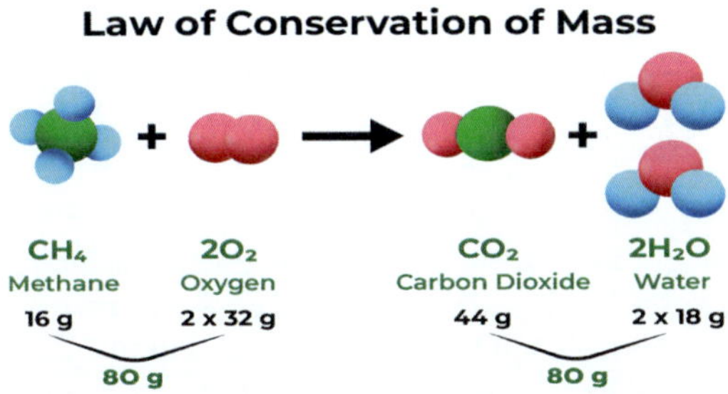

Applications of Law of Conservation of Mass

Law of Conservation of Mass is a fundamental principle that has many practical applications in various fields. Its understanding is crucial in the development of new technologies, the optimization of processes, and the protection of human health and the environment.

- **Chemical reactions:** The Law of Conservation of Mass is used to balance chemical equations, which is crucial in determining the stoichiometry of a reaction, the amount of reactants required, and the amount of products that will be formed. This law helps chemists to predict the outcome of a chemical reaction and to design and optimize chemical processes.
- **Environmental science:** The Law of Conservation of Mass is used to understand the fate and transport of pollutants in the environment. It helps environmental scientists to determine the sources and sinks of pollutants, to track their movement through ecosystems, and to assess the impacts of pollutants on human health and the environment.
- **Nuclear reactions:** The Law of Conservation of Mass is used to understand the principles of nuclear fission and fusion. It helps nuclear scientists to predict the amount of energy that will be released in a nuclear reaction and to design nuclear power plants and nuclear weapons.
- **Industrial processes:** The Law of Conservation of Mass is used to optimize industrial processes, such as chemical synthesis and manufacturing, and to minimize waste and resource consumption.
- **Material sciences:** The Law of Conservation of Mass is used to study the properties of materials and their behavior under different conditions. It helps material scientists to design new materials with specific properties and to optimize manufacturing processes to minimize waste and improve efficiency.

99. LAW OF EQUIPARTITION OF ENERGY

The Law of Equipartition of Energy is a fundamental principle in statistical mechanics that describes how the energy of a system is distributed among its degrees of freedom. It states that in thermal equilibrium, the energy is equally distributed among all of the degrees of freedom of the system, with each degree of freedom having an average energy of (1/2)kT, where k is the Boltzmann constant and T is the temperature of the system.

A degree of freedom is a way in which a particle in the system can store energy, such as through its kinetic energy, vibrational energy, or rotational energy. The Law of Equipartition of Energy applies to all systems in thermal equilibrium, including gases, liquids, and solids. The equipartition of kinetic energy was proposed initially in 1843, and more correctly in 1845, by John James Waterston. In 1859, James Clerk Maxwell argued that the kinetic heat energy of a gas is equally divided between linear and rotational energy. In 1876, Ludwig Boltzmann expanded on this principle by showing that the average energy was divided equally among all the independent components of motion in a system.

John Waterston

James Clerk Maxwell

Applications of Law of Equipartition of Energy

The Law of Equipartition of Energy has several important applications in physics and engineering, including:

- **Calculation of thermodynamic properties:** The Law of Equipartition of Energy is used to calculate the thermodynamic properties of a system, such as the specific heat capacity and the entropy. By knowing the number of degrees of freedom in a system, engineers and scientists can use the law to predict its thermodynamic behavior.
- **Understanding the behavior of materials:** The Law of Equipartition of Energy is used to understand the behavior of materials at different temperatures. At low temperatures, some degrees of freedom may be "frozen out," meaning that they no longer store energy. This can lead to changes in the material's properties, such as its thermal conductivity or specific heat capacity.
- **Study of chemical reactions:** The Law of Equipartition of Energy is used to study chemical reactions and the behavior of molecules. By understanding how energy is distributed among the degrees of freedom of a molecule, scientists can predict the rates of chemical reactions and the stability of molecules under different conditions.
- **Design of engines and machines:** The Law of Equipartition of Energy is used in the design of engines and machines that convert thermal energy into mechanical energy. By knowing the temperature and the number of degrees of freedom of the system, engineers can optimize the efficiency of the device.
- **Understanding the behavior of atoms and molecules:** The Law of Equipartition of Energy is used to understand the behavior of atoms and molecules in a variety of physical and chemical systems. For example, it can be used to predict the behavior of gases under different conditions, or to understand the dynamics of chemical reactions.

100. LAW OF RECIPROCAL PROPORTIONS

The law of reciprocal proportions, also known as the law of equivalent proportions, states that when two elements A and B combine with each other to form more than one compound, the weights of A that combine with a fixed weight of B bear a simple whole-number ratio to one another. In other words, the ratio of the masses of two elements that combine separately with the same amount of another element will be in the ratio of small whole numbers. This law was first stated by the French chemist Joseph Louis Proust in 1799. The law helped establish the concept of atomic masses and atomic proportions and played a significant role in the development of stoichiometry.

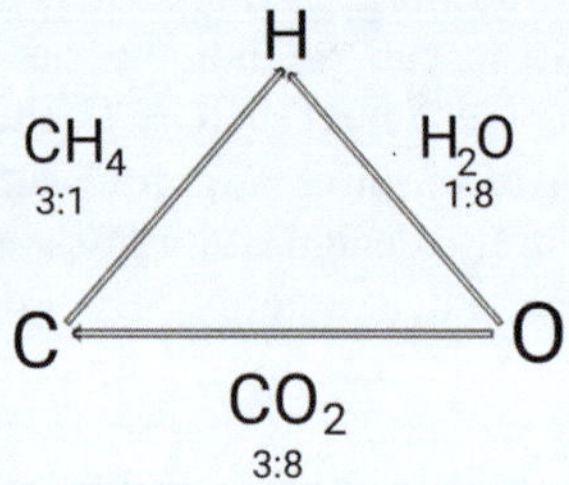

Joseph Louis Proust
(Credit: https://wellcomeimages.org/indexplus/image/L0006972.html)

Applications of Law of Reciprocal Proportions

The law of reciprocal proportions has many applications in chemistry.

- **Determining empirical formulas:** By knowing the relative weights of the elements in a compound, the empirical formula of the compound can be determined.
- **Predicting the products of reactions:** If the reactants are known, this law can be used to predict the ratios of the products formed.
- **Analyzing compounds:** The law can be used to analyze compounds to determine their composition and identify any impurities.
- **Synthesizing compounds:** The law of reciprocal proportions can be used to synthesize compounds with a desired composition and purity.

101. Law of Reflection

The law of reflection is a fundamental principle of optics that describes how light behaves when it reflects off a surface. It states that the angle of incidence (the angle between the incident light and the normal to the surface) is equal to the angle of reflection (the angle between the reflected light and the normal to the surface). In other words, the incident and reflected rays lie in the same plane, and the angle of incidence equals the angle of reflection.

This law can be stated mathematically as:

$$\theta_i = \theta_r$$

where θ_i is the angle of incidence, and θ_r is the angle of reflection.

The law of reflection applies to all types of waves, including light waves, sound waves, and water waves. It is the basis for many applications in optics, such as mirrors, lenses, and optical fibers.

The law of reflection can also be used to explain why objects appear in mirrors. The light that reflects off an object and enters the mirror follows the law of reflection and reflects back to the observer's eye, creating an image that appears to be behind the mirror.

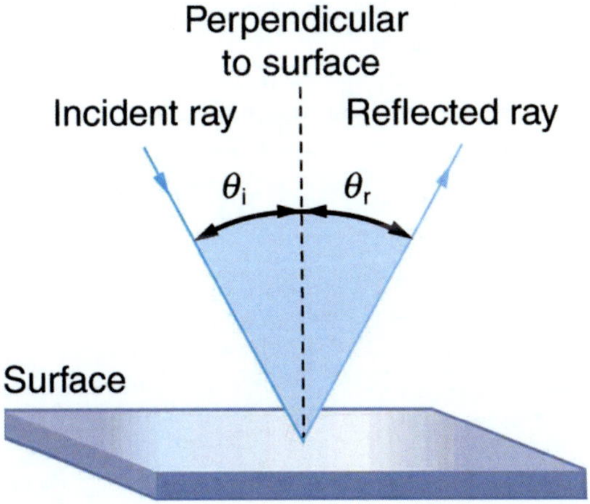

Applications of Law of Reflection

The law of reflection, which states that the angle of incidence of a wave is equal to the angle of reflection, has a number of applications in optics and related fields. Some notable applications include:

- **Mirrors:** The law of reflection is the fundamental principle behind how mirrors work. When light hits a flat mirror, the angle of incidence is equal to the angle of reflection, which allows us to see a reflection of the object in the mirror.
- **Periscopes:** Periscopes use multiple mirrors to reflect light and allow a person to see objects around corners or over obstacles. The law of reflection is used to ensure that the light is reflected at the correct angles to provide a clear view of the object.
- **Optical fibers:** Optical fibers are used to transmit light over long distances with minimal loss of signal. The fibers use the law of reflection to keep the light bouncing back and forth within the fiber, which reduces attenuation and ensures that the light arrives at the other end with minimal distortion.
- **Telescopes:** Telescopes use mirrors and lenses to collect and focus light from distant objects. The law of reflection is used to ensure that the light is reflected at the correct angles to create a clear image of the object.
- **Headlights:** The reflectors in car headlights use the law of reflection to focus the light from the bulb and direct it in a specific direction. By reflecting the light at a specific angle, the headlights can illuminate the road ahead without blinding other drivers.
- **Photography:** The law of reflection is important in photography, particularly in the use of reflectors to manipulate lighting. By reflecting light onto a subject from different angles, photographers can create different lighting effects and highlight specific areas of the subject.

102. LAW OF REFRACTION

The law of refraction, also known as Snell's law, describes how light bends or changes direction when it passes from one medium to another, such as when it passes through a lens or a prism. The law of refraction states that the ratio of the sine of the angle of incidence (the angle between the incident ray and the normal to the surface) to the sine of the angle of refraction (the angle between the refracted ray and the normal to the surface) is equal to the ratio of the velocities of light in the two media.

This law can be stated mathematically as:

$$n_1 \sin(\theta_1) = n_2 \sin(\theta_2)$$

where n_1 and n_2 are the refractive indices of the two media, θ_1 is the angle of incidence, and θ_2 is the angle of refraction.

The law of refraction explains why light bends when it passes through a prism or a lens, and why objects can appear distorted when viewed through a curved lens. It is also the basis for many optical applications, such as the design of lenses and the calculation of the critical angle for total internal reflection in optical fibers.

The law of refraction is an important principle in physics and optics, and it has many practical applications in fields such as medicine, telecommunications, and materials science.

Willebrord Snell

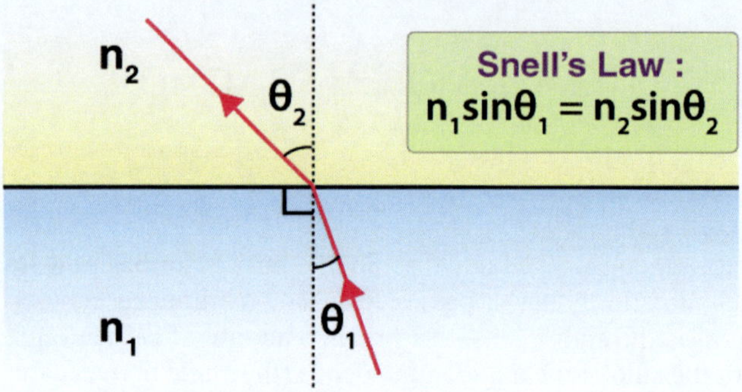

Applications of Law of Refraction

The law of refraction, also known as Snell's law, describes how light changes direction as it passes from one medium to another with a different refractive index. The law of refraction has numerous practical applications in various fields, including:

- **Lenses:** The law of refraction is fundamental to the design and operation of lenses, which are used in cameras, eyeglasses, telescopes, and other optical instruments. By bending light as it passes through a lens, the law of refraction helps to focus light onto a specific point or image.
- **Fiber optics:** The law of refraction is also crucial in the design of fiber optic cables, which are used to transmit data over long distances. By reflecting light within the fiber using the law of refraction, the cables can transmit data with minimal loss of signal.
- **Prisms:** Prisms use the law of refraction to split white light into its component colors, creating a rainbow effect. By refracting different wavelengths of light at different angles, prisms are used in a wide range of applications, including spectroscopy and optical communications.
- **Microscopy:** The law of refraction is used in microscopy to magnify and image small objects. By using a combination of lenses and the law of refraction, microscopes can magnify objects many times their original size, allowing for detailed observations of microorganisms, cells, and other small objects.
- **Mirage:** The law of refraction is responsible for creating mirages, which occur when light is bent as it passes through air of different densities, such as hot air above a road. The bending of light creates an optical illusion that appears as a distorted image or reflection.

103. LAWS OF FRICTION

The laws of friction describe the behavior of frictional forces between two surfaces in contact. These laws are:

- **The first law of friction:** Friction is proportional to the force pushing the surfaces together. In other words, the frictional force between two surfaces is directly proportional to the normal force pressing them together.
- **The second law of friction:** The force of friction opposes the direction of motion or the tendency of motion (or impending motion) between the two surfaces in contact. This means that the frictional force acts in the opposite direction to the applied force.
- **The third law of friction:** The force of friction is independent of the area of contact between the two surfaces. This means that the frictional force remains the same, regardless of the size of the contact area between the two surfaces.

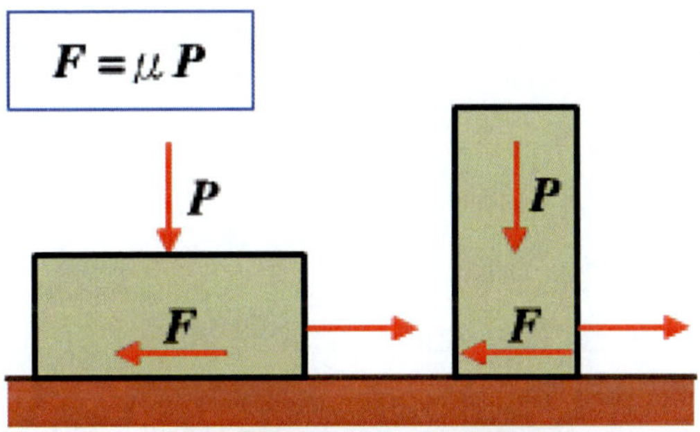

Coulomb's Laws of Friction
(Friction is proportional to the load pressing the solids together and almost independent of the area of contact.)

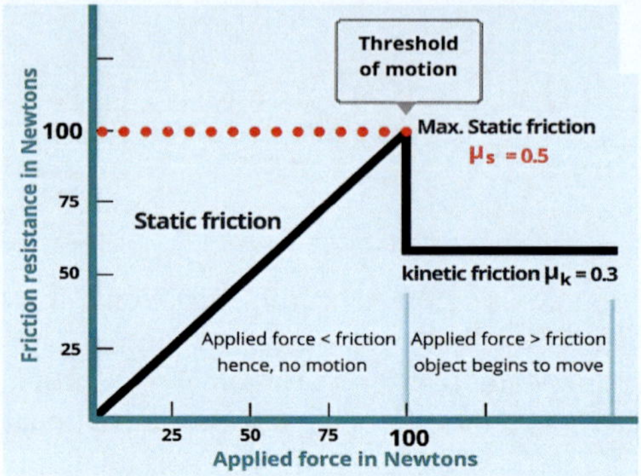

Applications of Laws of Friction

These laws have important implications in many fields, including engineering, physics, and materials science. They are essential for understanding the behavior of surfaces in contact, and for the design and optimization of various machines and structures. Here are some of the applications:

- **Design and optimization of machinery:** Engineers use the laws of friction to design and optimize machines and mechanisms that involve sliding, rolling, or static friction. For example, they use friction to control the motion of parts in a machine, and to prevent slipping or sliding of parts in contact.
- **Vehicle dynamics:** The laws of friction are used in the design of vehicles to control their stability, handling, and braking. For example, tire-road friction is critical in determining the braking distance and steering response of a vehicle.
- **Materials science:** Materials scientists study friction to better understand the behavior of materials under different conditions. For example, they study the friction between two surfaces in contact to better understand the wear and tear of materials, and to develop new materials with improved frictional properties.
- **Tribology:** Tribology is the study of friction, wear, and lubrication. It has many applications in various fields, including machine design, manufacturing, and maintenance.
- **Geology:** Geologists study the laws of friction to understand the behavior of rocks and soils under different conditions. For example, they study the friction between rocks in fault zones to better understand earthquakes and other geologic phenomena.

104. LEAVITT LAW

The Leavitt Law, also known as the Period-Luminosity Relation or the Cepheid Variable Period-Luminosity Relation, is an empirical relationship between the period of pulsation and the intrinsic luminosity (brightness) of certain types of variable stars known as Cepheid variables. The law was discovered by the Henrietta Swan Leavitt in the early 20th century.

Henrietta Leavitt's work involved studying the variable stars in the Small Magellanic Cloud, a satellite galaxy of the Milky Way. She examined a specific type of variable stars called Cepheid variables, which exhibit regular and predictable variations in their brightness over time.

Leavitt discovered a remarkable relationship between the period of pulsation and the intrinsic luminosity of Cepheid variables. She found that the longer the period of variation, the more intrinsically luminous the star. This period-luminosity relationship enabled astronomers to estimate the distances to these stars, as the apparent brightness of a star is affected by both its intrinsic luminosity and its distance from Earth.

The Leavitt Law revolutionized the field of astronomy by providing a reliable means to measure distances to galaxies beyond the Milky Way.

Henrietta Leavitt's discovery of the period-luminosity relationship for Cepheid variables is considered a fundamental contribution to the field of astronomy and has had a profound impact on our understanding of the universe.

Henrietta Swan Leavitt

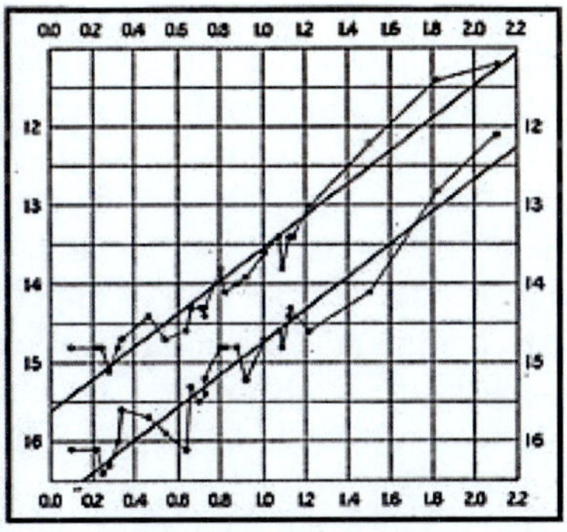

Period-Luminosity relation

(Leavitt and Pickering 1912, 2)

Applications of Leavitt Law

The Leavitt Law, or the period-luminosity relationship for Cepheid variables, has had several important applications in the field of astronomy. Here are some notable applications:

- **Distance measurements:** The Leavitt Law provides a reliable means to estimate distances to galaxies beyond the Milky Way. By observing the period of pulsation of Cepheid variables in distant galaxies and comparing them with known Cepheids closer to us, astronomers can calculate the distances to those galaxies.
- **Cosmological distance scale:** The Leavitt Law has played a crucial role in constructing the cosmological distance scale, which relates the redshift of galaxies (a measure of their recessional velocity due to the expansion of the universe) to their distances. By calibrating the Leavitt Law with nearby galaxies and applying it to more distant ones, astronomers can estimate the distances to far-off galaxies and map the expansion history of the universe.
- **Determination of the Hubble constant:** The Hubble constant describes the rate at which the universe is expanding. The Leavitt Law, combined with observations of Cepheid variables in nearby galaxies, has been used to calibrate the relationship between their period and luminosity, providing a crucial tool for measuring the Hubble constant. Determining the Hubble constant helps refine our understanding of the age and evolution of the universe.
- **Stellar population studies:** The Leavitt Law is useful for studying the properties and evolution of stellar populations within galaxies. By observing Cepheid variables and their periods in different galaxies, astronomers can investigate the ages, compositions, and evolutionary stages of stellar populations in various environments.
- **Standard candles for cosmology:** Cepheid variables serve as "standard candles" in cosmology—objects with known intrinsic luminosities. By applying the Leavitt Law, astronomers can determine the distances to Cepheid variables, which in turn can be used to calibrate other distance indicators, such as Type I_a supernovae.
- **Extragalactic astrophysics:** The Leavitt Law has broader implications in the study of extragalactic astrophysics. Astronomers can explore various astrophysical phenomena, such as the properties of active galactic nuclei, star formation rates in galaxies, the evolution of galaxy clusters, and the distribution of dark matter in the universe.

105. Le Chatelier's Principle

Le Chatelier's principle is a fundamental concept in chemistry that describes how a system at equilibrium responds to changes in external conditions such as temperature, pressure, or concentration of reactants or products. The principle states that a system at equilibrium will shift its equilibrium position in response to a change in one of these external factors, in such a way as to partially counteract the change and restore equilibrium.

For example, consider a chemical reaction where reactants A and B combine to form product C. If the concentration of A is increased, Le Chatelier's principle predicts that the equilibrium will shift to the right, favoring the formation of more product C, until a new equilibrium is reached. Similarly, if the temperature of the reaction is increased, the principle predicts that the equilibrium will shift in the direction that absorbs heat, either to the left or right, depending on whether the reaction is exothermic or endothermic.

Le Chatelier's principle provides a powerful tool for predicting the behavior of chemical systems under different conditions, and it is used extensively in chemical engineering and process design to optimize reaction conditions and maximize product yield. However, it is important to note that the principle is a qualitative rather than a quantitative guideline, and that the actual response of a system to a perturbation may depend on a variety of other factors such as kinetics, thermodynamics, and molecular structure.

Henry Louis Le Chatelier

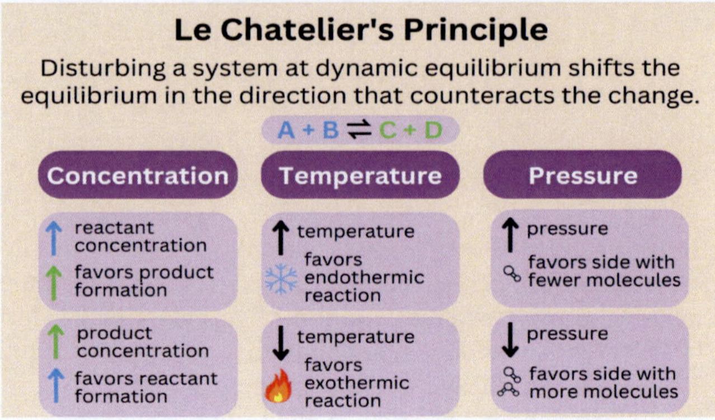

Applications of Le Chatelier's Principle

Le Chatelier's principle is a fundamental concept in chemistry that states that a system at equilibrium will respond to any stress or change in conditions by shifting its equilibrium position in a way that partially counteracts the effect of the change. This principle has numerous practical applications in various fields, including:

- **Chemical reactions:** Le Chatelier's principle is used to predict the direction in which a chemical reaction will shift when conditions such as temperature, pressure, or concentration are changed. This allows chemists to optimize reaction conditions and yields, as well as to design new reactions.
- **Industrial processes:** The principle is used in industrial processes such as the production of ammonia, where changes in temperature and pressure can be used to optimize the yield of the desired product.
- **Environmental chemistry:** Le Chatelier's principle is also relevant in environmental chemistry, where it is used to predict how ecosystems will respond to changes in environmental conditions such as pH, temperature, and nutrient concentrations.
- **Biological systems:** The principle is also relevant in biological systems, where it is used to understand how enzymes and other biomolecules respond to changes in temperature, pH, and other conditions.
- **Electrochemistry:** Le Chatelier's principle is used to predict the direction of electron flow in electrochemical cells when conditions such as temperature, concentration, or pressure are changed.

106. LEIDENFROST EFFECT

The Leidenfrost effect is a phenomenon in which a liquid, when brought into contact with a surface much hotter than its boiling point, forms a layer of vapor that insulates it from the surface and slows down the rate of evaporation. This effect was first described by a German scientist Johann Gottlob Leidenfrost in 1756.

When a liquid is brought into contact with a hot surface, it immediately begins to evaporate. However, when the temperature of the surface is much higher than the boiling point of the liquid, a thin layer of vapor is formed between the liquid and the surface. This layer of vapor reduces the contact area between the liquid and the surface, thus reducing the rate of heat transfer from the surface to the liquid. As a result, the liquid can remain in a superheated state for a longer period of time.

The Leidenfrost effect has many practical applications, such as in cooking, where it is used to determine the temperature of a hot pan by observing the behavior of water droplets on its surface, and in industry, where it is used in metalworking and forging processes to prevent the liquid coolant from boiling away too quickly.

Gottlob Leidenfrost

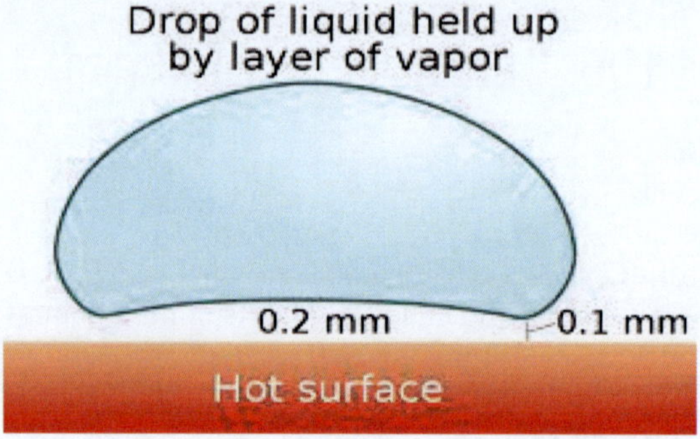

Applications of Leidenfrost Effect

The Leidenfrost effect is a phenomenon that occurs when a liquid comes into contact with a surface that is significantly hotter than the liquid's boiling point. The liquid will then form a thin layer of vapor that insulates it from the hot surface, resulting in a characteristic "dancing droplet" effect. Here are some notable applications of the Leidenfrost effect:

- **Cooling:** The Leidenfrost effect can be used for cooling applications, such as in heat exchangers or refrigeration systems. By using a surface that is significantly hotter than the liquid, the Leidenfrost effect can provide efficient cooling with minimal energy input.
- **Fire extinguishing:** The Leidenfrost effect can be used to extinguish fires by spraying a mist of water onto a fire. The water droplets will vaporize and form a layer of vapor that insulates the fire from oxygen, effectively extinguishing it.
- **Energy harvesting:** The Leidenfrost effect can be used to harvest energy from waste heat, such as from industrial processes or engines. By using a surface that is significantly hotter than the liquid, the Leidenfrost effect can create a temperature difference that can be used to generate electricity.
- **Self-cleaning surfaces:** The Leidenfrost effect can be used to create self-cleaning surfaces by using a hydrophobic coating that promotes the formation of a Leidenfrost layer. The layer of vapor can help to repel dirt and other contaminants, keeping the surface clean.
- **Microfluidics:** The Leidenfrost effect can be used to manipulate liquids in microfluidic devices, such as by using the dancing droplet effect to move and mix small volumes of liquid.

107. Lenz's Law

Lenz's law, named after the Russian physicist Heinrich Lenz, is a fundamental principle in electromagnetism. It describes the direction of induced current in a conductor when it is exposed to a changing magnetic field. Lenz's Law states that an induced current in a conductor will always flow in such a direction as to oppose the change that produced it. In other words, when there is a change in the magnetic field through a conductor, an electromotive force (EMF) is induced in the conductor, and this EMF creates an electric current that produces a magnetic field that opposes the original change.

This law is a consequence of the conservation of energy and the principle of electromagnetic induction, which was discovered by Michael Faraday.

Heinrich Friedrich Lenz

Michael Faraday

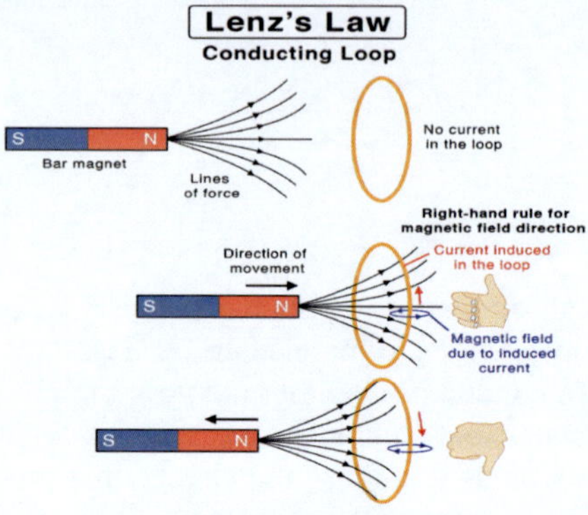

Applications of Lenz's Law

Lenz's Law is important in many areas of physics and engineering, including electrical generators, transformers, motors, and inductors. Some practical applications of Lenz's Law include:

- **Eddy current brakes:** Lenz's Law is used in the design of eddy current brakes, which are used in trains and other vehicles to slow them down. These brakes work by inducing a current in a conductor as it moves through a magnetic field, and the resulting magnetic field opposes the motion of the conductor, producing a braking force.
- **Induction heating:** Lenz's Law is also used in induction heating, which is a process that uses electromagnetic induction to heat metal objects. When a metal object is placed in a varying magnetic field, an eddy current is induced in the object, which produces heat due to the resistance of the metal to the current.
- **Transformers:** Lenz's Law is important in the design of transformers, which are devices that transfer electrical energy from one circuit to another through electromagnetic induction. The law ensures that the induced current in the secondary coil of the transformer opposes the change in the magnetic field produced by the primary coil, which allows for efficient energy transfer.
- **Electric generators:** Lenz's Law is also important in electric generators, which convert mechanical energy into electrical energy. As the rotor of a generator spins, it creates a changing magnetic field, which induces an EMF in the stator coil, producing an electric current. The law ensures that the direction of the induced current opposes the change in the magnetic field, which helps to produce a steady output voltage.

108. Leonard–Merritt Mass Estimator

The Leonard-Merritt mass estimator is a formula used in astronomy to estimate the mass of a galaxy cluster based on the observed velocities of its member galaxies. It is named after the astronomers A. Leonard and D. Merritt, who developed the formula in the 1980s.

The Leonard-Merritt mass estimator is based on the assumption that the cluster is in dynamical equilibrium, meaning that the gravitational forces between the galaxies are balanced by their motions. The formula relates the total mass of the cluster to the velocity dispersion of its member galaxies, which is a measure of the spread of their velocities around the average value. Specifically, the formula is:

$$M = - 3 R \sigma^2 / G$$

where, M is the estimated mass of the galaxy cluster. R is the radius of the cluster. σ is the velocity dispersion of the member galaxies. G is the gravitational constant.

It should be noted that the formula is based on several simplifying assumptions, such as spherical symmetry and isotropy of the velocity dispersion, which may not be satisfied in all cases. Furthermore, the estimator relies on accurate measurements of the velocities of a large number of galaxies, which can be challenging to obtain in practice.

David Roy Merritt

(Credit: https://academictree.org/physics/peopleinfo.php?pid=250550)

Applications of Leonard–Merritt Mass Estimator

Leonard-Merritt mass estimator is a valuable tool in astronomy for estimating the masses of galaxies and other astronomical objects based on their observed velocities. It has numerous applications in studying galactic dynamics, black holes, and the distribution of matter in the universe, and can also be used to test alternative theories of gravity. The method has several practical applications, including:

- **Measuring the masses of galaxies:** The Leonard-Merritt mass estimator is used to estimate the masses of galaxies and their dark matter halos. This allows astronomers to better understand the distribution of matter in the universe and the evolution of galaxies over time.
- **Determining the masses of black holes:** The method can also be used to estimate the mass of a black hole based on the motion of stars or gas around it. This has been used to study supermassive black holes at the centers of galaxies.
- **Studying galactic dynamics:** The Leonard-Merritt mass estimator is also used to study the dynamics of galaxies, including their rotation curves, velocity dispersions, and other properties.
- **Testing theories of gravity:** The method has also been used to test alternative theories of gravity, such as modified Newtonian dynamics (MOND), which attempt to explain the observed motion of galaxies without invoking dark matter.

109. Lorentz Force Law

The Lorentz force law describes the force exerted on a charged particle in an electromagnetic field. It is named after the Dutch physicist Hendrik Lorentz, who first derived the equation in 1895.

The Lorentz force law states that the force F on a particle with charge q moving with velocity v in an electromagnetic field with electric field E and magnetic field B is given by:

$$\mathbf{F} = \mathbf{q}\ (\mathbf{E} + \mathbf{v} \times \mathbf{B})$$

where × denotes the cross product. The electric field E exerts a force on the charged particle proportional to its charge, while the magnetic field B exerts a force perpendicular to the direction of motion of the charged particle, depending on the direction of the magnetic field and the velocity of the particle.

The Lorentz force law is fundamental to the understanding of the behavior of charged particles in magnetic and electric fields, and has important applications in a wide range of fields, including particle accelerators, plasma physics, and electromechanical devices.

Hendrik Lorentz

$$\vec{F} = q\vec{E} + q\vec{v} \times \vec{B}$$

Electric force *Magnetic force*

Applications of Lorentz Force Law

The Lorentz force law is a fundamental principle in electromagnetism with numerous practical applications in many fields, including electric motors and generators, particle accelerators, magnetic resonance imaging, electromagnetic levitation, and plasma physics. The law has numerous practical applications in various fields, including:

- **Electromagnetic motors and generators:** The Lorentz force law is the basis for the operation of electric motors and generators, which convert electrical energy into mechanical energy and vice versa. The force on a current-carrying wire in a magnetic field produces torque, causing the rotor of a motor or generator to rotate.
- **Particle accelerators:** The Lorentz force law is used to design and operate particle accelerators, such as synchrotrons and cyclotrons, which use electromagnetic fields to accelerate charged particles to high energies.
- **Magnetic resonance imaging (MRI):** MRI machines use strong magnetic fields to image the body and diagnose medical conditions. The Lorentz force law plays a key role in the operation of MRI machines, as it is used to manipulate the spin of protons in the body's tissues.
- **Electromagnetic levitation:** The Lorentz force law is also used in the design of electromagnetic levitation systems, which use magnetic fields to levitate and stabilize objects without physical contact. This technology is used in magnetic levitation trains, maglev roller coasters, and other transportation systems.
- **Plasma physics:** The Lorentz force law is used to describe the behavior of charged particles in plasmas, which are ionized gases that are used in many industrial processes, including fusion energy research.

110. Lorentz Transformations

Lorentz transformations are a set of equations that describe how physical measurements of time and space are affected by motion at high speeds, close to the speed of light. They were first formulated by Dutch physicist Hendrik Lorentz in the late 19th century and later refined by Albert Einstein in his theory of special relativity.

Lorentz transformations are expressed mathematically using four-dimensional space-time, where time and space are treated as a single entity. The transformations involve the concepts of time dilation (the slowing down of time for an observer in motion relative to a stationary observer) and length contraction (the shortening of objects in the direction of motion).

The equations for Lorentz transformations involve the speed of light (c), the relative velocity of two observers (v), and the coordinates of events in space-time (t, x, y, and z). The transformations preserve the speed of light as a constant and ensure that the laws of physics are the same for all observers, regardless of their relative motion.

The Lorentz transformations have many applications in modern physics, including the understanding of high-energy particle physics, the behavior of electromagnetic waves, and the GPS system. They are also essential for the development of modern technologies such as particle accelerators and nuclear reactors.

Hendrik Lorentz

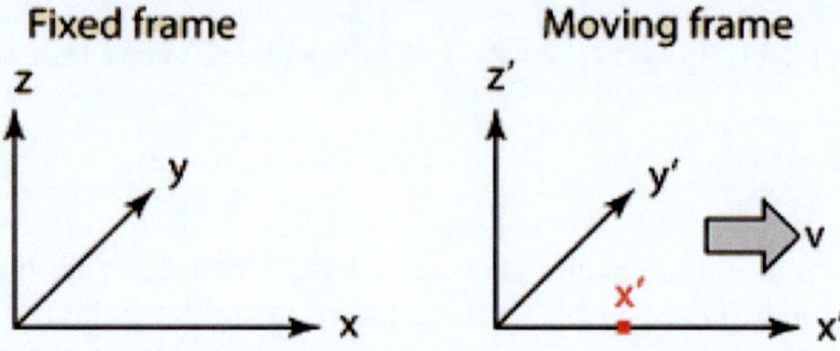

Applications of Lorentz Transformations

The Lorentz transformations are a set of mathematical equations that describe how space and time coordinates are transformed between different reference frames in special relativity. Lorentz transformations have many practical applications in fields such as particle physics, GPS, aviation, and satellite communications, including:

- **Relativistic particle physics:** The Lorentz transformations are essential for understanding the behavior of high-energy particles, such as those produced in particle accelerators. The equations are used to calculate the relativistic energy and momentum of these particles, which are critical for predicting their behavior and interactions.
- **GPS:** The Global Positioning System (GPS) is a network of satellites that use the principles of special relativity, including the Lorentz transformations, to determine the precise location of objects on Earth. The satellites carry atomic clocks that are synchronized with clocks on the ground, and the GPS receiver uses the time difference between the clocks to calculate the distance between the satellite and the receiver.
- **Time dilation:** One of the most well-known effects of special relativity is time dilation, which occurs when an object moves at a significant fraction of the speed of light. The Lorentz transformations are used to calculate the amount of time dilation that occurs in different reference frames, which has practical applications in fields such as aviation and satellite communications.
- **Relativistic optics:** The Lorentz transformations are also used to describe the behavior of light in special relativity. For example, the equations can be used to calculate the relativistic Doppler effect, which describes the frequency shift of light emitted by a moving object.

111. MACH NUMBER

The Mach number is a dimensionless quantity that describes the speed of an object moving through a fluid, relative to the speed of sound in that fluid. It is named after Austrian physicist and philosopher Ernst Mach.

The Mach number is defined as:

Mach number = velocity of object / speed of sound

When the Mach number is less than one, the object is traveling at subsonic speeds, meaning it is moving slower than the speed of sound. When the Mach number is equal to one, the object is traveling at the speed of sound, which is also known as the sonic speed. When the Mach number is greater than one, the object is traveling at supersonic speeds, meaning it is moving faster than the speed of sound.

The Mach number is an important parameter in aerodynamics, where it is used to describe the behavior of fluids and objects moving through them. It is particularly important in the design of aircraft and rockets, where the behavior of the fluid flow around the object at different Mach numbers is critical to their performance.

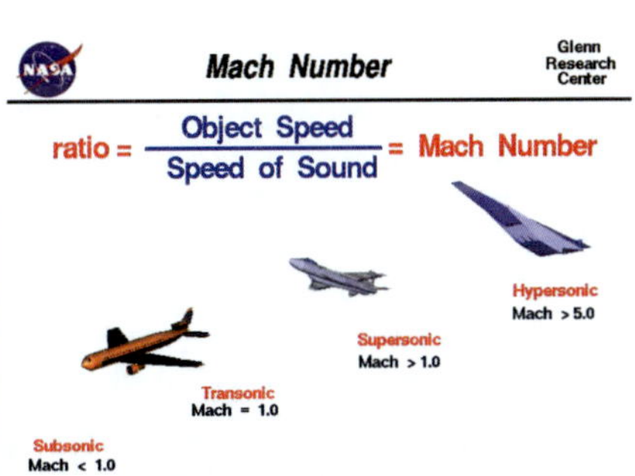

(a) Ernst Mach **(b) Mach Number**

(Credit (a): https://wellcomeimages.org/indexplus/image/L0023068.html)
(Credit (b): https://www.grc.nasa.gov/www/BGH/mach.html)

Applications of Mach Number

Mach number is a critical parameter in the study of fluid dynamics and is used extensively in the design of aircraft, rockets, and other objects that move through fluids. Its applications range from aircraft design to wind tunnel testing, and it is an important parameter in understanding the behavior of fluids at different speeds. Major applications are:

- **Aircraft design:** The Mach number is used extensively in the design of aircraft, where it is critical to ensure that the aircraft can operate effectively at the different speeds it will encounter. Designing an aircraft that can operate at both subsonic and supersonic speeds requires careful consideration of the fluid dynamics involved, which can be described using the Mach number.
- **Rocket design:** The Mach number is also important in the design of rockets, where it is critical to ensure that the rocket can operate effectively as it moves through the atmosphere and into space. The Mach number can be used to calculate the optimal shape and size of the rocket, as well as the amount of fuel needed to achieve the desired speed.
- **Sonic boom:** When an object moves faster than the speed of sound, it creates a shock wave known as a sonic boom. The Mach number is used to calculate the intensity of the sonic boom, which is important in understanding its effects on people and structures on the ground.
- **Wind tunnel testing:** Wind tunnel testing is a common technique used to study the behavior of objects moving through fluids. The Mach number is used to control the speed of the fluid in the wind tunnel, allowing researchers to study the behavior of the object at different speeds.
- **Sonic booms:** Sonic booms occur when an object moves through the air at supersonic speeds, creating a shock wave. The intensity of the sonic boom is related to the Mach number of the object, and it is an important parameter in the study of sonic booms and their effects on the environment.
- **High-speed trains:** High-speed trains, such as the Japanese Shinkansen or the French TGV, operate at speeds that can be close to the speed of sound. The Mach number is an important parameter in the design and operation of these trains, and it is used to optimize the aerodynamic properties of the train, as well as to minimize noise and vibration.
- **Hypersonic vehicles:** Hypersonic vehicles, such as spaceplanes or missiles, operate at speeds greater than five times the speed of sound. The Mach number is a critical parameter in the design and operation of these vehicles, and it is used to optimize the vehicle's shape, propulsion, and thermal protection system.

112. MACH PRINCIPLE

The Mach principle is a hypothesis proposed by the physicist and philosopher Ernst Mach in the 19th century. According to this principle, the properties of matter are determined by their interaction with the rest of the universe. Specifically, the inertia of an object is related to the distribution of matter in the universe.

At its core, Mach's principle suggests that the inertia of an object is influenced by the distribution of all the other matter in the universe. In other words, the motion of an object is not solely determined by its own properties but is also influenced by the mass and distribution of all other objects in the cosmos.

Mach argued that an object's inertial mass arises from its interaction with all the other matter in the universe, which affects the object's motion. He believed that the universe's matter distribution determines the frame of reference in which the laws of physics appear to be valid. In other words, the inertial properties of matter depend on the distribution of matter in the universe as a whole, rather than just the local environment.

Mach's ideas were influential in the development of relativity theory, particularly in the work of Albert Einstein. However, the Mach principle is not considered a fundamental law of physics, and its role in shaping our understanding of the universe remains a topic of debate among physicists and philosophers.

Ernst Mach

(Credit: https://academictree.org/physics/peopleinfo.php?pid=44533)

Applications of Mach Principle

The Mach principle is a concept in physics that was developed by the Austrian physicist Ernst Mach. Mach principle has had significant theoretical applications in the fields of general relativity, cosmology, and quantum mechanics. While the principle has not been experimentally verified, it has provided important insights into our understanding of the universe and the nature of gravity. It suggests that the inertial properties of an object are determined by the distribution of matter in the universe. The principle has several theoretical applications, including:

- **General relativity:** The Mach principle was an important influence on the development of Einstein's theory of general relativity. The principle suggests that the force of gravity is not an inherent property of massive objects, but rather is a result of the distribution of matter in the universe. This idea helped Einstein to develop his theory of gravitation, which describes how massive objects curve space-time and influence the motion of other objects.
- **Cosmology:** The Mach principle has implications for our understanding of the structure and evolution of the universe. The principle suggests that the universe is not a static, isolated system, but rather is a dynamic and interconnected system in which the distribution of matter affects the motion of all objects. This idea has been used to explain various observations, such as the large-scale structure of the universe and the motion of galaxies.
- **Quantum mechanics:** The Mach principle has also been applied to the field of quantum mechanics. In particular, it has been used to explain the phenomenon of quantum entanglement, in which two particles can be correlated in such a way that their properties are dependent on each other, even at great distances. The Mach principle suggests that the correlation between these particles arises from the distribution of matter in the universe.

113. MACH REFLECTION

Mach reflection is a phenomenon that occurs when a shock wave, such as a blast wave from an explosion, encounters a reflecting surface. It is named after the Austrian physicist and philosopher Ernst Mach, who first described it.

In a typical scenario, a shock wave travels through a medium and encounters a solid surface at an angle. The shock wave is partially reflected from the surface and partially transmitted through it. If the angle of incidence is large enough, the reflected and transmitted waves can intersect, creating a complex pattern of shock waves known as a Mach stem.

The Mach stem is characterized by a sudden increase in pressure and temperature behind the reflected shock wave. This can cause the shock wave to become more intense and to spread out in a fan-shaped pattern. As a result, the region of high pressure behind the shock wave becomes much larger than it would be in the absence of the reflecting surface.

Mach reflection is an important phenomenon in the study of explosions, as it can have significant effects on the damage caused by the blast wave. It is also of interest in the study of supersonic flows and high-speed aerodynamics, as similar phenomena can occur when shock waves encounter obstacles or curved surfaces.

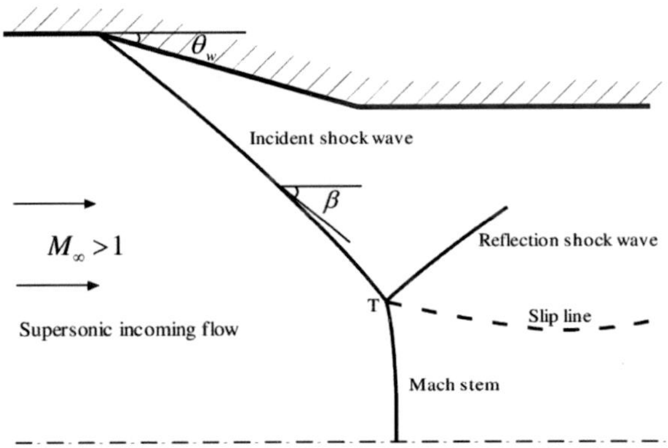

Applications of Mach Reflection

Mach reflection is a phenomenon in fluid dynamics that occurs when a shock wave reflects back onto itself. The reflected wave creates a region of high pressure and temperature behind the shock wave, known as the Mach stem. Mach reflection has important applications in various fields, including aerospace engineering, explosive engineering, and geophysics. Understanding the behavior of shock waves and Mach reflection is essential for designing efficient and effective systems, predicting natural disasters, and advancing our understanding of fluid dynamics. Mach reflection has several applications in various fields, including:

- **Aerospace engineering:** Mach reflection is an important consideration in the design of supersonic aircraft and rockets. The formation of a Mach stem can lead to significant changes in the flow field and can affect the performance and stability of the vehicle. Engineers use computational fluid dynamics simulations to model the behavior of shock waves and Mach reflection in order to optimize the design of supersonic vehicles.
- **Explosive engineering:** Mach reflection is also important in the design and analysis of explosive devices, such as bombs and missiles. The formation of a Mach stem can increase the pressure and temperature in the blast wave, leading to more damage and destruction. Explosive engineers use computational models to predict the behavior of shock waves and Mach reflection in different environments in order to design effective and efficient explosive devices.
- **Geophysics:** Mach reflection can also occur in natural phenomena such as earthquakes and volcanic eruptions. The reflection of shock waves in the Earth's crust can lead to the formation of secondary waves, which can have significant effects on the ground motion and seismic activity. Geophysicists study the behavior of shock waves and Mach reflection in order to better understand the dynamics of earthquakes and volcanic eruptions and to improve their predictive capabilities.

114. MALUS' LAW

Malus' law, also known as Malus-Duplex law, is a fundamental principle of optics that describes the behavior of light waves that pass through a polarizing filter. It was discovered by French physicist Étienne-Louis Malus in 1808.

The law states that the intensity of light transmitted through a polarizing filter is proportional to the square of the cosine of the angle between the transmission axis of the filter and the polarization direction of the incident light. Mathematically, Malus' law is expressed as:

$$I = I_0 \cos^2\theta$$

where I is the intensity of the transmitted light, I_0 is the maximum intensity of the incident light, θ is the angle between the transmission axis of the filter and the polarization direction of the incident light.

Malus' law finds applications in various fields such as optics, polarimetry, and astronomy. It is also used in the production of sunglasses, 3D glasses, and other optical devices that use polarizing filters.

Étienne-Louis Malus

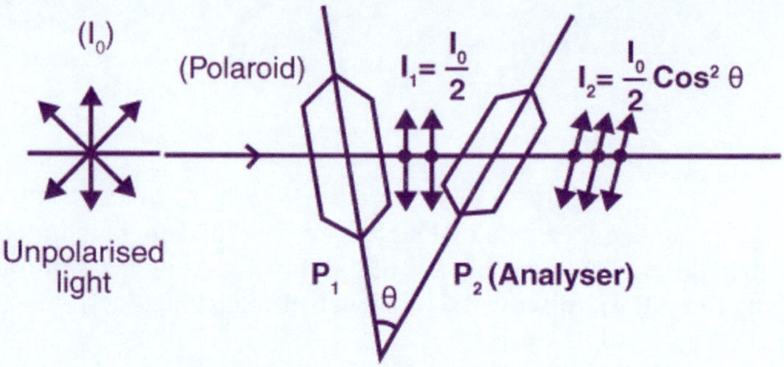

Applications of Malus' Law

Malus' law, also known as the cosine law, describes the polarization of light when it passes through a polarizer. Malus' law has important applications in a wide range of fields, including optics, materials science, and remote sensing. By understanding the behavior of polarized light, scientists and engineers can develop new technologies and applications that rely on the polarization of light waves.

- **Polarized sunglasses:** Malus' law is used in the design and manufacturing of polarized sunglasses. The lenses of polarized sunglasses contain a polarizing filter that blocks out horizontal light waves, reducing glare and improving visibility in bright sunlight.
- **LCD displays:** Liquid crystal displays (LCDs) use polarizers to control the orientation of light waves passing through the display. Malus' law is used to calculate the amount of light transmitted through the polarizer, which is then used to adjust the brightness and contrast of the display.
- **Optical microscopy:** Malus' law is used in optical microscopy to determine the orientation of polarized light waves passing through a sample. By analyzing the intensity of polarized light passing through the sample at different angles, researchers can obtain information about the molecular structure and composition of the sample.
- **Remote sensing:** Malus' law is used in remote sensing applications to analyze the polarization of light reflected from the Earth's surface. By measuring the polarization of light at different angles, scientists can obtain information about the physical properties of the Earth's surface, such as the presence of water, vegetation, and mineral deposits.

115. Marconi's Law

Marconi's law refers to the relationship between the height of an antenna and the maximum communication distance of a radio transmission. It was formulated by the Italian physicist and inventor Guglielmo Marconi in the early 20th century. Marconi's law states that the maximum distance between two wireless communication points is proportional to the height of the transmitting antenna multiplied by the height of the receiving antenna. The law assumes that both antennas are placed above the surface of the Earth, with no obstructions between them. The relationship is based on the fact that the curvature of the Earth limits the range of radio signals, but increasing the height of the antennas can overcome this limitation. Marconi's law was an important factor in the development of radio communication and remains relevant today in the design of modern communication systems.

Guglielmo Marconi

Applications of Marconi's Law

Marconi's work had a profound impact on the development of modern communication and navigation technologies. His pioneering work paved the way for many of the technologies we rely on today, making it easier for people around the world to connect and communicate with each other. Guglielmo Marconi was an Italian inventor and electrical engineer who is known for his pioneering work in the development of radio communication. Here are some applications of Marconi's work:

- **Wireless communication:** Marconi's most significant contribution to science and technology was the invention of the first practical wireless communication system. His work paved the way for the development of modern wireless communication technologies such as cell phones, satellite communication, and Wi-Fi.
- **Radio broadcasting:** Marconi's invention of the radio receiver made it possible for people to listen to radio broadcasts from all over the world. Today, radio remains an important medium of communication and entertainment, providing news, music, and other programming to millions of people around the globe.
- **Radar:** Marconi's work on wireless communication also laid the groundwork for the development of radar technology. Radar is used in a wide range of applications, from air traffic control and weather forecasting to military surveillance and missile defense.
- **Navigation:** Marconi's invention of wireless communication systems also had important applications in navigation, allowing ships and planes to communicate with each other and with ground-based stations. Today, GPS and other satellite-based navigation systems rely on wireless communication technology to provide accurate location information to users.

116. Mass-Energy Equivalence

Mass-energy equivalence is a concept in physics that states that mass and energy are interchangeable and can be converted from one form to another. This principle is encapsulated by Albert Einstein's famous equation,

$$E = m\,c^2$$

where E represents energy, m mass, and c the speed of light in a vacuum.

According to the equation, a small amount of mass can be converted into a large amount of energy, and vice versa, with the conversion factor being the speed of light squared. This means that even a tiny amount of mass contains an enormous amount of potential energy.

The most famous example of mass-energy equivalence is the process of nuclear reactions, where a small amount of mass is converted into a tremendous amount of energy. Nuclear reactions, such as those occurring in the Sun or in nuclear power plants, involve the conversion of a small fraction of the mass of atomic nuclei into energy.

Another example is the production of energy in particle accelerators, where high-speed particles collide and release energy. The Large Hadron Collider (LHC) is an example of a particle accelerator where the mass-energy equivalence is used to study fundamental particles and their interactions. Mass-energy equivalence has been experimentally confirmed through various experiments, such as the observation of the energy released in nuclear reactions or the annihilation of particles and antiparticles, where their mass is converted entirely into energy.

The concept of mass-energy equivalence is a fundamental principle in physics and has important implications in fields such as nuclear physics, particle physics, and cosmology.

Albert Einstein

Applications of Mass-Energy Equivalence

The concept of mass-energy equivalence has numerous applications. Here are some notable applications:

- **Nuclear Energy:** The most significant application of mass-energy equivalence is in the field of nuclear energy. Nuclear power plants generate electricity by harnessing the energy released from nuclear reactions, such as nuclear fission (splitting of heavy atomic nuclei) or nuclear fusion (combining light atomic nuclei). These reactions convert a small amount of mass into a significant amount of energy.
- **Particle Physics:** Mass-energy equivalence plays a crucial role in particle physics research, particularly in particle accelerators. Accelerators like the Large Hadron Collider (LHC) accelerate particles to nearly the speed of light and collide them together. When particles collide, some of their mass is converted into energy, creating new particles. By studying these particle interactions and the energy released, scientists can explore the fundamental properties of matter and the nature of the universe.
- **Nuclear Weapons:** Mass-energy equivalence is central to the destructive power of nuclear weapons. Atomic bombs rely on the release of a large amount of energy by converting a small amount of mass into energy through nuclear fission or fusion. The energy release in a nuclear explosion is enormous and can cause devastating effects.
- **Medical Applications:** Mass-energy equivalence is used in medical imaging techniques such as positron emission tomography (PET) scans. PET scans involve the detection of gamma rays emitted by positron-emitting radionuclides. When positrons (antiparticles of electrons) collide with electrons, they annihilate, converting their combined mass into energy in the form of gamma rays. This energy is detected and used to create images of the body, helping in the diagnosis and monitoring of various medical conditions.
- **Space Travel:** Mass-energy equivalence is relevant to space exploration and propulsion. Concepts like nuclear propulsion and antimatter propulsion rely on the conversion of mass into energy to achieve high velocities and overcome the limitations of conventional chemical rockets. While these technologies are still in the realm of theoretical research, they demonstrate the potential for harnessing mass-energy equivalence for advanced space travel.

The principle has far-reaching implications in many other areas of physics, such as cosmology, quantum field theory, and the study of black holes, where it helps us understand the nature of the universe and the interactions of matter and energy at fundamental scales.

117. MAXWELL'S EQUATIONS

Maxwell's equations are a set of four partial differential equations that describe the behavior of electric and magnetic fields in space, and their relationship to charges and currents. They were first formulated by James Clerk Maxwell in the 1860s and are the foundation of classical electromagnetism.

The four equations are:

- **Gauss's law for electric fields:** The electric flux through any closed surface is proportional to the charge enclosed within the surface. Mathematically, this is written as

$$\nabla \cdot \mathbf{E} = \rho / \varepsilon_0$$

where $\nabla \cdot E$ is the divergence of the electric field E, ρ is the charge density, and ε_0 is the electric constant (also known as the permittivity of free space).

- **Gauss's law for magnetic fields:** The magnetic flux through any closed surface is zero. Mathematically, this is written as

$$\nabla \cdot \mathbf{B} = 0$$

where $\nabla \cdot B$ is the divergence of the magnetic field B.

- **Faraday's law of induction:** A changing magnetic field induces an electric field. Mathematically, this is written as

$$\nabla \times \mathbf{E} = - d\mathbf{B}/dt$$

where $\nabla \times E$ is the curl of the electric field E, and dB/dt is the time derivative of the magnetic field B.

- **Ampere's law with Maxwell's correction:** A changing electric field produces a magnetic field, in addition to the magnetic field produced by a current. Mathematically, this is written as

$$\nabla \times \mathbf{B} = \mu_0 (\mathbf{J} + \varepsilon_0 (d\mathbf{E}/dt))$$

where $\nabla \times B$ is the curl of the magnetic field B, μ_0 is the magnetic constant (also known as the permeability of free space), J is the current density, and dE/dt is the time derivative of the electric field E.

Together, these four equations describe the behavior of electric and magnetic fields and their interaction with charges and currents in space.

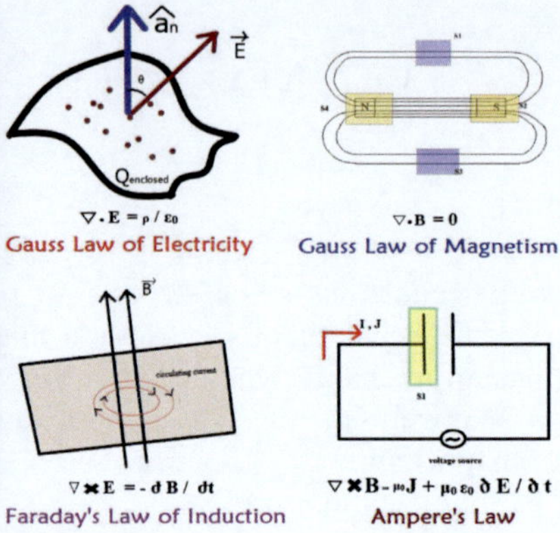

James Clerk Maxwell
(Credit: https://academictree.org/physics/peopleinfo.php?pid=12313)

Applications of Maxwell's Equations

Maxwell's equations are a fundamental set of equations that describe the behavior of electromagnetic fields. They have numerous practical applications in telecommunications, electrical engineering, medical imaging, and astronomy, among other fields.

- **Telecommunications:** The equations are essential for understanding the behavior of electromagnetic waves, which are used in telecommunications for transmitting and receiving information.
- **Electrical engineering:** The equations are used in the design of electrical circuits and devices such as transformers, motors, and generators.
- **Medical imaging:** Maxwell's equations are used in medical imaging techniques such as magnetic resonance imaging (MRI), which relies on the behavior of magnetic fields in the human body.
- **Astronomy:** The equations are used in the study of astrophysical phenomena such as black holes, neutron stars, and pulsars, which involve strong electromagnetic fields.

118. MAXWELL'S RELATIONS

Maxwell's relations are a set of four equations in thermodynamics that relate the partial derivatives of thermodynamic variables. These relations are named after the Scottish physicist James Clerk Maxwell who derived them in the 19th century. Maxwell's relations can be stated as:

Maxwell relation for internal energy (U):

$$(\partial T/\partial V)_S = - (\partial P/\partial S)_V$$

Maxwell relation for enthalpy (H):

$$(\partial T/\partial P)_S = (\partial V/\partial S)_P$$

Maxwell relation for Helmholtz free energy (A):

$$(\partial S/\partial V)_T = (\partial P/\partial T)_V$$

Maxwell relation for Gibbs free energy (G):

$$(\partial S/\partial P)_T = - (\partial V/\partial T)_P$$

In these relations, S represents entropy, V represents volume, P represents pressure, and T represents temperature. The left-hand side of each equation represents a partial derivative of a thermodynamic property with respect to one variable, while the right-hand side represents a partial derivative of another property with respect to a different variable.

The Maxwell relations are important in thermodynamics because they allow one to calculate a thermodynamic variable by knowing the values of other variables. For example, if the pressure-volume relationship of a gas is known, the Maxwell relations can be used to determine the heat capacity or the change in volume with respect to temperature. These relations are also used in the study of phase transitions and in the analysis of thermodynamic cycles.

James Clerk Maxwell

The four principal thermodynamic energies, their differential expressions, and the corresponding Maxwell relations.

Thermodynamic energy	Differential expression	Corresponding Maxwell relations
U	$dU = TdS - PdV$	$\left(\frac{\partial T}{\partial V}\right)_S = -\left(\frac{\partial P}{\partial S}\right)_V$
H	$dH = TdS + VdP$	$\left(\frac{\partial T}{\partial P}\right)_S = \left(\frac{\partial V}{\partial S}\right)_P$
A	$dA = -SdT - PdV$	$\left(\frac{\partial S}{\partial V}\right)_T = \left(\frac{\partial P}{\partial T}\right)_V$
G	$dG = -SdT + VdP$	$\left(\frac{\partial S}{\partial P}\right)_T = -\left(\frac{\partial V}{\partial T}\right)_P$

Applications of Maxwell's Relations

The Maxwell relations have a wide range of applications in various fields of science and engineering, including in the study of materials, chemical reactions, and biological systems. They are particularly useful in the design and analysis of energy conversion systems, such as engines, refrigeration cycles, and heat pumps. Maxwell's relations are a fundamental concept in thermodynamics, and their applications are extensive and diverse. Some of the applications of Maxwell's relations are:

- **Thermodynamics:** Maxwell's relations are fundamental in the study of thermodynamics. They allow one to relate the different thermodynamic variables and calculate their values based on known variables.
- **Chemical reactions:** Maxwell's relations are used in the study of chemical reactions and their kinetics. By knowing the enthalpy, entropy, and free energy of a reaction, Maxwell's relations can be used to determine the temperature and pressure conditions required for the reaction to occur.
- **Phase transitions:** Maxwell's relations are used to study phase transitions in materials. By knowing the thermodynamic variables, such as the enthalpy, entropy, and free energy, Maxwell's relations can be used to determine the conditions under which a phase transition occurs and the critical points at which the transition occurs.
- **Materials science:** Maxwell's relations are used in the development of new materials, such as alloys and polymers. By knowing the thermodynamic properties of the materials, Maxwell's relations can be used to determine the optimal processing conditions required to produce the desired material properties.
- **Engineering:** Maxwell's relations are used in the design and analysis of energy conversion systems, such as engines, refrigeration cycles, and heat pumps. They are also used in the design of heat exchangers, which are used in many industrial processes.
- **Biological systems:** Maxwell's relations are used in the study of biological systems, such as metabolism and cellular respiration. By knowing the thermodynamic properties of the biological molecules involved, Maxwell's relations can be used to determine the conditions required for the metabolic reactions to occur.

119. MIKHEYEV–SMIRNOV–WOLFENSTEIN EFFECT

The Mikheyev-Smirnov-Wolfenstein (MSW) effect is a phenomenon in neutrino oscillations, which explains the modification of the neutrino's oscillation probabilities as they travel through matter. The effect is named after its discoverers, Stanislav Mikheyev, Alexei Smirnov, and Lincoln Wolfenstein, who independently proposed it in the early 1980s.

The MSW effect is caused by the interaction of the neutrinos with matter in the form of electrons and nuclei. As a neutrino travels through matter, it may interact with the electrons and/or nuclei, leading to a change in its effective mass and thus modifying its oscillation behavior. The effect is most pronounced for neutrinos with low energies, such as those produced in the sun.

The MSW effect has been observed in numerous experiments, including solar neutrino experiments, atmospheric neutrino experiments, and reactor neutrino experiments. Its discovery was a major breakthrough in our understanding of neutrino physics and helped to establish the idea that neutrinos have nonzero masses.

Stanislav Mikheyev **Alexei Smirnov**

(http://nuclphys.sinp.msu.ru/persons/pages/mikheyev_stanislav.htm)
(http://nuclphys.sinp.msu.ru/persons/pages/smirnov_alexei.htm)

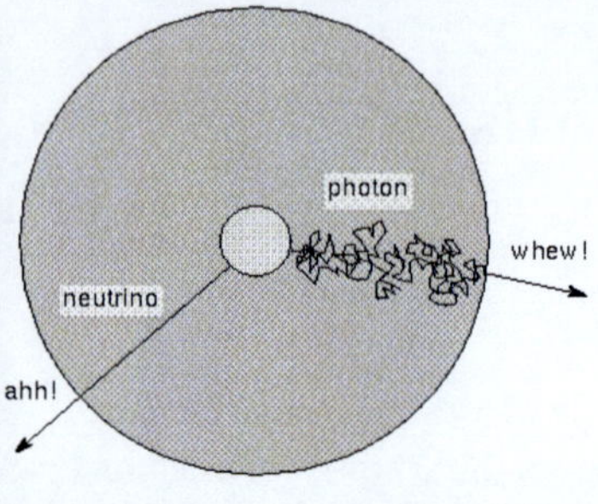

Photons take tortuous paths out of the Sun's interior. Neutrinos pass right on through in just two seconds.

(a) Lincoln Wolfenstein (b) Neutrino

(Credit (a): https://academictree.org/physics/peopleinfo.php?pid=126339)
(b): (Strobel, Nick. 2022. "Neutrino". Accessed June 19, 2022)

Applications of Mikheyev–Smirnov–Wolfenstein Effect

The Mikheyev-Smirnov-Wolfenstein (MSW) effect is a phenomenon in neutrino oscillations that occurs when neutrinos pass through matter, causing them to change flavor as they travel. MSW effect has important implications for our understanding of the properties of neutrinos and the structure of the universe. Here are some applications of the MSW effect:

- **Solar neutrinos:** The MSW effect is responsible for the observed deficit in the flux of electron neutrinos from the Sun. As solar neutrinos pass through the dense matter in the Sun, they undergo flavor oscillations, leading to a reduction in the number of electron neutrinos detected on Earth.
- **Neutrino detection:** The MSW effect can be used to enhance the detection of neutrinos in experiments. By using detectors that are sensitive to specific neutrino flavors, scientists can use the MSW effect to enhance the signal of certain types of neutrinos while suppressing others.
- **Neutrino tomography:** The MSW effect can be used to study the structure and composition of the Earth's interior by using neutrinos as probes. By measuring the flavor composition of neutrinos that have passed through the Earth, scientists can infer the density and composition of different layers of the Earth's interior.
- **Dark matter detection:** The MSW effect can be used to search for dark matter particles by looking for neutrinos produced in the annihilation of dark matter particles in the Sun. The MSW effect can enhance the detection of these neutrinos, providing a potential indirect detection method for dark matter.

120. Mitscherlich's Law

Mitscherlich's law, also known as the law of isomorphism, states that crystals with similar chemical composition tend to have similar crystal structures. This means that if the crystal structure of one substance is known, it is possible to predict the crystal structure of another substance with a similar chemical composition.

Mitscherlich's law was first proposed by Eilhardt Mitscherlich, a German chemist, in 1819. It was based on his observation that several compounds with the same chemical composition, such as sodium chloride and potassium chloride, had similar crystal structures. He concluded that the crystal structure of a substance was related to its chemical composition.

Mitscherlich's law is an important principle in crystallography, as it allows scientists to predict the crystal structures of new materials based on their chemical composition. This is useful in many fields, including materials science, chemistry, and geology.

Eilhardt Mitscherlich

ISOMORPHISM

The term isomorphism is applied to the phenomenon of substances with analogous formulas having closely related crystal structures. The term was introduced by Mitscherlich in 1819, who prepared crystals of KH_2PO_4, KH_2AsO_4, $(NH_4)H_2PO_4$ and $(NH_4)H_2AsO_4$ and found that they showed the same forms and the interfacial angles between corresponding faces were very similar.

Applications of Mitscherlich's Law

Mitscherlich's law is a principle in crystallography that states that the shape of a crystal is determined by the shape of the unit cell. It has several applications in materials science and engineering:

- **Crystal structure determination:** Mitscherlich's law is used to determine the crystal structure of materials. By analyzing the symmetry and shape of the crystal unit cell, scientists can determine the arrangement of atoms in the crystal lattice.
- **Crystal growth:** The law is used in crystal growth technology to control the shape and size of crystals. By controlling the shape of the unit cell, crystal growers can produce crystals with desired properties and shapes.
- **Properties of materials:** The law is also used to predict the properties of materials. For example, the electrical conductivity of a crystal can be predicted based on the shape of the unit cell.
- **Powder X-ray diffraction:** The law is used in powder X-ray diffraction, a technique used to determine the crystal structure of materials. The technique involves measuring the diffraction pattern of X-rays that have been scattered by the crystal lattice. The pattern can be analyzed to determine the shape and size of the unit cell.
- **Thin film growth:** Mitscherlich's law is also used in the growth of thin films, which are used in a wide range of electronic and optical devices. By controlling the shape of the unit cell, scientists can grow thin films with specific properties, such as high conductivity or high transparency.

121. MOORE'S LAW

Moore's Law is a prediction made by Gordon Moore, the co-founder of Intel, in 1965 that the number of transistors on a microchip would double about every two years, resulting in an exponential increase in computing power. This law has been remarkably accurate over the years and has driven the growth of the semiconductor industry, leading to the development of faster and more powerful computers and mobile devices. However, it is worth noting that in recent years, the law has faced some challenges due to the physical limitations of semiconductor technology, and some experts believe that it may not hold true in the future.

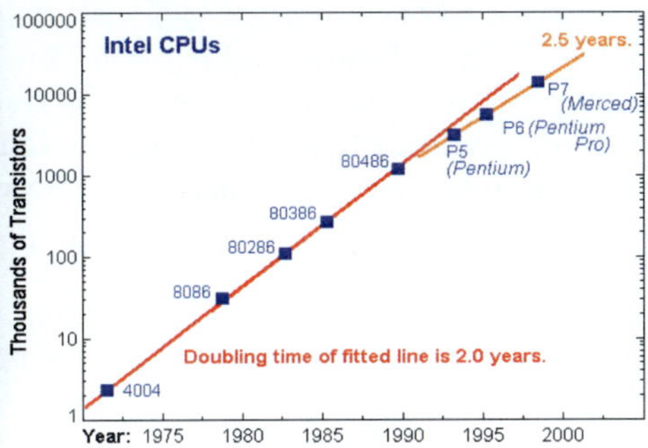

Gordon Moore
(Credit: https://www.flickr.com/photos/oninnovation/4334769385)

Applications of Moore's Law

Moore's law is an observation made by Gordon Moore in 1965, which states that the number of transistors on a microchip doubles approximately every two years, resulting in a rapid increase in computing power and decrease in the cost of technology. The applications of Moore's law are widespread and have had a profound impact on technology and society. Here are some examples:

- **Increased processing power:** As the number of transistors on a chip increases, so does the processing power of computers. This has enabled the development of complex software and applications that require significant computing power, such as artificial intelligence, machine learning, and data analysis.

- **Miniaturization of devices:** The shrinking size of transistors has allowed for the development of smaller, more portable devices such as laptops, smartphones, and wearable technology.
- **Lower costs:** The decreasing cost of microchips has made technology more accessible to a wider range of people and businesses, allowing for the development of new industries and applications.
- **Advancements in medical technology:** Moore's law has facilitated the development of advanced medical technologies such as MRI machines, DNA sequencing, and medical imaging equipment.
- **Increased automation:** The increased processing power has enabled the development of robots and automated systems that can perform tasks more efficiently and accurately than humans.
- **Improved communication:** The widespread use of the internet, social media, and mobile devices is due in part to the increasing power and decreasing cost of microchips.
- **Environmental impact:** The increasing efficiency of technology has the potential to reduce energy consumption and greenhouse gas emissions, leading to a more sustainable future.

Moore's law has had a profound impact on technology and society, driving innovation and shaping the way we live and work. However, it is important to note that there are physical limits to the number of transistors that can be placed on a chip, and researchers are exploring new technologies to continue the trend of increasing computing power.

122. Moseley's Law

Moseley's law, also known as the Moseley's law of X-ray spectra, states that the square root of the frequency of the characteristic X-ray emission of an element is proportional to its atomic number. This law was discovered by the English physicist Henry Moseley in 1913. Mathematically, Moseley's law can be expressed as:

$$\sqrt{(\nu)} = K (Z - \sigma)$$

where $\sqrt{(\nu)}$ represents the square root of the X-ray frequency, Z is the atomic number of the element, K is a constant, and σ is an empirical parameter that may vary for different elements.

Moseley's law is significant because it provided the first experimental evidence for the concept of atomic number, which is the number of protons in the nucleus of an atom. Prior to Moseley's discovery, the elements were arranged in the periodic table based on their atomic weights, which sometimes led to inconsistencies and inaccuracies in predicting the properties of unknown elements.

Moseley's law helped to establish a more accurate and consistent periodic table based on atomic number. It also provided a way to determine the atomic number of an unknown element by analyzing its X-ray spectra.

Henry Moseley

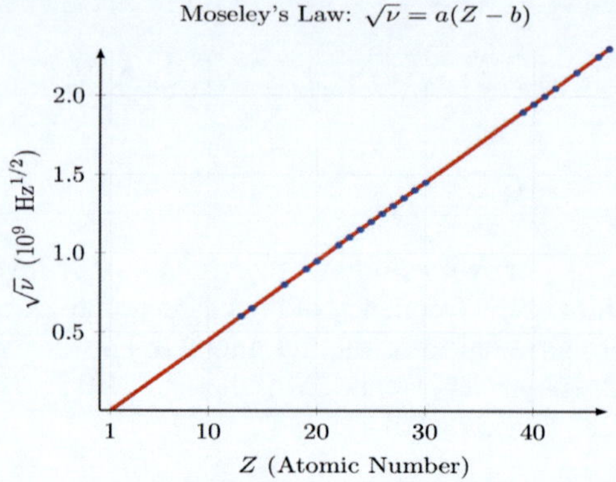

Applications of Moseley's Law

Moseley's law has several important applications in physics and chemistry, including:

♦ **Determination of atomic number:** Moseley's law provides a way to determine the atomic number of an unknown element by analyzing its characteristic X-ray spectra. This technique is widely used in analytical chemistry and materials science.

♦ **Study of atomic structure:** Moseley's law provides important information about the structure of atoms and their electronic configuration. It has helped to refine our understanding of the arrangement of electrons in atoms and their energy levels.

♦ **Prediction of chemical properties:** Moseley's law is used in predicting the chemical properties of unknown elements and their compounds. By knowing the atomic number of an element, it is possible to predict its chemical behavior and reactivity.

♦ **Development of X-ray spectroscopy:** Moseley's law played a critical role in the development of X-ray spectroscopy, which is a powerful tool for studying the structure of matter at the atomic and molecular level. X-ray spectroscopy is widely used in physics, chemistry, and materials science.

♦ **Verification of quantum mechanics:** Moseley's law provided important experimental evidence for the validity of the quantum mechanical model of the atom. It showed that the energies of atomic electrons are quantized and that the quantum numbers used to describe the electronic configuration of atoms have a physical significance.

123. Murphy's Law

Murphy's Law is a popular adage that states "Anything that can go wrong, will go wrong". The law is often invoked to express the idea that if something can fail, it is likely to do so at the worst possible moment. The law is named after Edward A. Murphy Jr., an American aerospace engineer who was involved in testing experimental equipment for the United States Air Force in the late 1940s.

The origins of Murphy's Law are somewhat disputed, but it is believed to have arisen from an incident in which a technician improperly installed a sensor during a rocket sled test, causing the test to fail. Murphy reportedly commented that "If there are two or more ways to do something, and one of those ways can result in a catastrophe, then someone will do it".

The concept of Murphy's Law has become a widely recognized cultural trope and has been used in various contexts, such as engineering, business, and personal life. Some examples of how Murphy's Law might apply in various situations include:

A business proposal that seems foolproof can still fail due to unforeseen circumstances.

A computer system that has been working perfectly for years can crash at the worst possible moment.

A carefully planned vacation can be ruined by unexpected weather, illness, or other setbacks.

A traffic jam that causes a person to be late for an important meeting can occur on the one day they are running late.

Edward A. Murphy Jr.

(Credit: https://www.daytoninnovationlegacy.org/murphy.html)

Applications of Murphy's Law

Despite its humorous and often fatalistic tone, Murphy's Law can serve as a reminder to plan for contingencies and to remain vigilant for potential problems in any situation.

- **Risk management:** Understanding Murphy's Law is an important aspect of risk management. It helps individuals and organizations to anticipate potential problems and plan accordingly. By considering all possible scenarios, they can be better prepared for unexpected events.
- **Engineering and product design:** Engineers and designers often use Murphy's Law to anticipate potential failures or malfunctions in their products. They conduct thorough testing and quality control measures to minimize the risk of failure.
- **Project management:** Murphy's Law is also relevant to project management. It encourages project managers to anticipate potential problems and risks and develop contingency plans to mitigate them. By doing so, they can avoid delays, unexpected costs, and other negative consequences.
- **Aviation and aerospace industry:** Murphy's Law is particularly relevant in the aviation and aerospace industry. Engineers and technicians use it to identify potential safety hazards and design systems to minimize the risk of accidents.
- **Personal development:** Murphy's Law can be applied to personal development as well. By anticipating potential obstacles and setbacks, individuals can develop resilience and prepare themselves to handle challenges effectively. It can also help individuals to cultivate a positive mindset and focus on solutions rather than problems.

Murphy's Law has wide-ranging applications in various fields, from risk management to personal development. By understanding and applying this principle, individuals and organizations can be better prepared to handle unexpected events and minimize the risk of negative consequences.

124. NAVIER–STOKES EQUATIONS

The Navier-Stokes equations describe the motion of fluid substances such as liquids and gases. They are named after French physicist Claude-Louis Navier and British mathematician and physicist George Gabriel Stokes, who first derived them in the 19th century.

The Navier-Stokes equations are a set of partial differential equations that describe the conservation of momentum and mass for a fluid. They express the fundamental relationship between the motion of a fluid and the forces acting on it, such as pressure, viscosity, and external forces.

The equations are given in vector form and consist of the continuity equation, which expresses the conservation of mass, and the momentum equation, which expresses the conservation of momentum. The momentum equation is further broken down into two separate equations for conservation of linear and angular momentum.

The Navier-Stokes equations are notoriously difficult to solve analytically, and even numerical solutions can be complex and computationally expensive. They are of great importance in fluid dynamics, and their solutions have applications in fields such as engineering, meteorology, and oceanography.

Claude-Louis Navier

George Gabriel Stokes

NAVIER-STOKES EQUATION

$$\rho g_x - \frac{\partial p}{\partial x} + \mu\left(\frac{\partial^2 u}{\partial x^2} + \frac{\partial^2 u}{\partial y^2} + \frac{\partial^2 u}{\partial z^2}\right) = \rho\frac{Du}{Dt}$$

$$\rho g_y - \frac{\partial p}{\partial y} + \mu\left(\frac{\partial^2 v}{\partial x^2} + \frac{\partial^2 v}{\partial y^2} + \frac{\partial^2 v}{\partial z^2}\right) = \rho\frac{Dv}{Dt}$$

$$\rho g_z - \frac{\partial p}{\partial z} + \mu\left(\frac{\partial^2 w}{\partial x^2} + \frac{\partial^2 w}{\partial y^2} + \frac{\partial^2 w}{\partial z^2}\right) = \rho\frac{Dw}{Dt}$$

Navier-Stokes Equation

$$\frac{\partial \rho}{\partial t} + \nabla\cdot(\rho\vec{V}) = 0$$

Continuity

Applications of Navier–Stokes Equations

The Navier-Stokes equations describe the behavior of fluids and are used extensively in the field of fluid dynamics. Navier-Stokes equations have many practical applications in a wide range of fields, from aerospace to medicine, and they continue to be an active area of research. Here are some applications of Navier-Stokes equations:

♦ **Aerospace:** The Navier-Stokes equations are used in the design of aircraft and spacecraft. They help engineers understand the behavior of air and fluids around these vehicles, and this understanding is critical for designing more efficient and safer aircraft.

♦ **Weather forecasting:** Numerical weather prediction models rely heavily on the Navier-Stokes equations to model the behavior of the atmosphere. These models help meteorologists predict the weather with greater accuracy.

♦ **Oceanography:** The Navier-Stokes equations are also used in the study of ocean currents and waves. Understanding the behavior of fluids in the ocean is important for marine transportation, offshore oil exploration, and climate modeling.

♦ **Industrial processes:** The Navier-Stokes equations are used in the design and optimization of industrial processes that involve the movement of fluids. This includes the design of pipelines, pumps, and turbines.

♦ **Medical applications:** The Navier-Stokes equations are used to study blood flow and other fluid flows in the human body. This understanding is critical for designing medical devices, such as artificial heart valves, and for developing treatments for cardiovascular diseases.

.

125. NERNST EQUATION

The Nernst equation is a fundamental equation in electrochemistry that describes the relationship between the potential difference (or voltage) and the concentration gradient of ions across a membrane or at the interface of two solutions. It is named after the German physical chemist Walther Nernst, who developed the equation in 1889.

The Nernst equation is given by:

$$E = E^{\circ} - (RT/nF) \ln(Q)$$

where, E is the cell potential (in volts)

E° is the standard cell potential (in volts)

R is the gas constant (8.314 J/mol.K)

T is the absolute temperature (in Kelvin)

n is the number of electrons involved in the reaction

F is the Faraday constant (96,485 C/mol)

Q is the reaction quotient, which is the ratio of the concentrations of the products to the concentrations of the reactants

The Nernst equation is used to calculate the equilibrium potential of an electrode, which is the voltage that develops across an electrode when there is no net flow of current. It is also used to determine the concentrations of ions in a solution, given the electrode potential. The Nernst equation has many applications in electrochemistry, including the measurement of pH, the calculation of equilibrium constants, and the determination of thermodynamic properties of electrochemical systems.

Walther Nernst

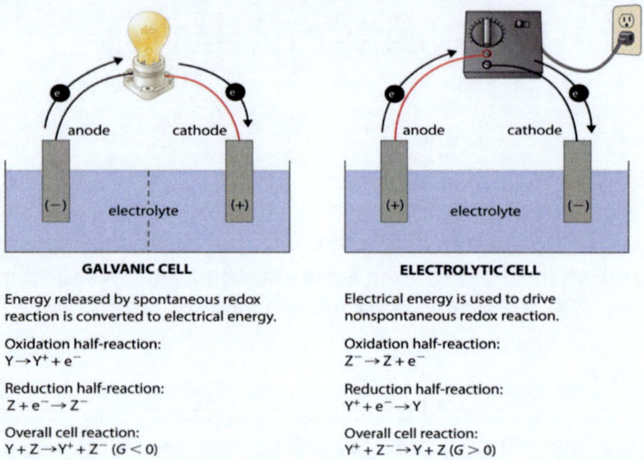

Applications of Nernst Equation

The Nernst equation relates the standard cell potential of an electrochemical reaction to the concentrations of the reactants and products and the temperature. Its applications are widespread in chemistry, biology, and engineering. Some examples are:

- **Electrochemistry:** The Nernst equation is a fundamental equation used to calculate the equilibrium potential of an electrochemical cell. It is used to determine the voltage at which a redox reaction reaches equilibrium and to calculate the standard cell potential.
- **Analytical chemistry:** The Nernst equation is used in analytical chemistry to measure the concentration of ions in solution. For example, in pH measurement, the Nernst equation is used to relate the concentration of hydrogen ions to the pH of a solution.
- **Biology:** The Nernst equation is used to calculate the membrane potential of a cell. It is also used to understand how ions move across cell membranes and to study the electrical properties of neurons.
- **Environmental engineering:** The Nernst equation is used in the design and operation of wastewater treatment plants. It is used to calculate the equilibrium concentration of ions in a wastewater treatment reactor and to optimize the operation of the reactor.
- **Corrosion:** The Nernst equation is used to predict the corrosion potential of metals in different environments. It is used to understand the corrosion behavior of metals and to develop strategies for corrosion prevention and control.

126. Newton's Law of Cooling

Newton's law of cooling states that the rate of heat loss of a body is directly proportional to the difference in temperature between the body and its surroundings. Mathematically, it can be expressed as:

$$dT/dt = -k \times (T - T_a)$$

where,

dT/dt is the rate of change of temperature with respect to time,

T is the temperature of the object,

T_a is the ambient temperature (temperature of the surrounding),

k is the cooling constant.

This law is commonly used in the fields of thermodynamics and heat transfer, and has applications in areas such as cooling of electronic devices, refrigeration, and air conditioning.

Isaac Newton

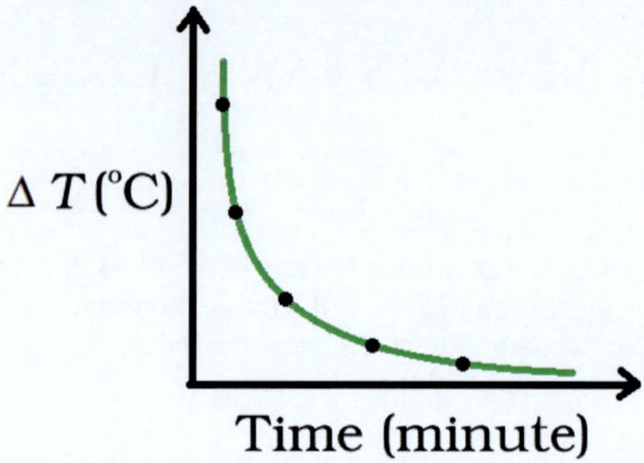

Applications of Newton's Law of Cooling

Newton's law of cooling is used in various fields, including physics, chemistry, engineering, and biology. Here are some specific applications:

- **Temperature measurement:** Newton's law of cooling can be used to measure the temperature of a substance by measuring its cooling rate over time.
- **Heat transfer:** The law is used in the study of heat transfer, which is important in many engineering applications, such as designing cooling systems and heat exchangers.
- **Food preservation:** The cooling of food products is essential for their preservation, and Newton's law can be used to calculate the cooling rate of a food product, which helps to determine the optimal cooling conditions for storage.
- **Climate science:** The law is used in the study of climate science, specifically in determining the cooling rate of the Earth's atmosphere.
- **Biomedical applications:** The cooling of living tissues and organs is important in biomedical applications, such as organ preservation for transplantation, and the law can be used to study the cooling rate of these tissues.
- **Forensic science:** Newton's law of cooling can be used to estimate the time of death of a person by analyzing the cooling rate of the body.
- **Materials science:** The law is used in the study of material properties, such as thermal conductivity, specific heat, and thermal diffusivity, which are important in materials science and engineering.

127. NEWTON'S LAW OF UNIVERSAL GRAVITATION

Newton's law of universal gravitation is a fundamental law of physics that describes the force of gravitational attraction between two objects. The law states that every particle in the universe attracts every other particle with a force that is proportional to the product of their masses and inversely proportional to the square of the distance between them. The mathematical expression of the law is:

$$F = G \times (m_1 \times m_2) / r^2$$

where F is the force of gravitational attraction between two objects, G is the gravitational constant, m_1 and m_2 are the masses of the two objects, and r is the distance between the centers of the two objects.

Isaac Newton

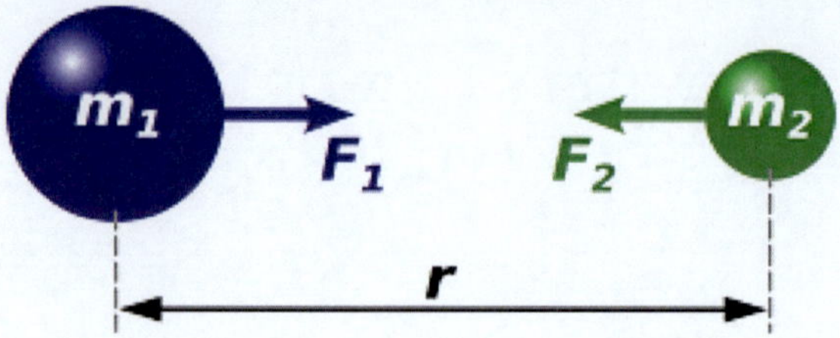

$$F_1 = F_2 = G\,\frac{m_1 \times m_2}{r^2}$$

Applications of Newton's Law of Universal Gravitation

Some of the practical applications of Newton's law of universal gravitation are:

- **Planetary motion:** Newton's law of universal gravitation is the foundation of our understanding of the motion of planets, moons, and other celestial bodies in the solar system. It helps us to predict the orbits of these bodies and to understand the forces that govern their motion.
- **Satellite motion:** The law of gravitation is also crucial in understanding the motion of artificial satellites, such as those used for communication, navigation, and weather monitoring. It helps to calculate the orbital velocities and trajectories of these satellites and to predict their positions at different times.
- **Gravity-related phenomena:** The law of gravitation explains many other gravity-related phenomena, such as tides, the motion of objects in free fall, and the behavior of objects in orbit around each other.
- **Gravitational lensing:** The bending of light by massive objects is another consequence of the law of gravitation, known as gravitational lensing. This phenomenon has practical applications in astronomy, such as the observation of distant galaxies and the detection of exoplanets.

Newton's law of universal gravitation is a fundamental law of physics that has numerous practical applications in astronomy, space exploration, and other fields. It is essential for understanding the motion of celestial bodies and artificial satellites, as well as many other gravity-related phenomena.

128. Newton's Law of Viscosity

Newton's Law of Viscosity, also known as the Law of Dynamic Viscosity, is a fundamental concept in fluid mechanics that describes the behavior of fluids when subjected to shear stress. The law states that the shear stress between layers of a fluid is proportional to the rate of change of velocity of those layers.

The formula for Newton's Law of Viscosity is as follows:

$$\tau = \eta\ (du / dy)$$

where τ is the shear stress, η is the dynamic viscosity, du/dy is the velocity gradient.

Isaac Newton

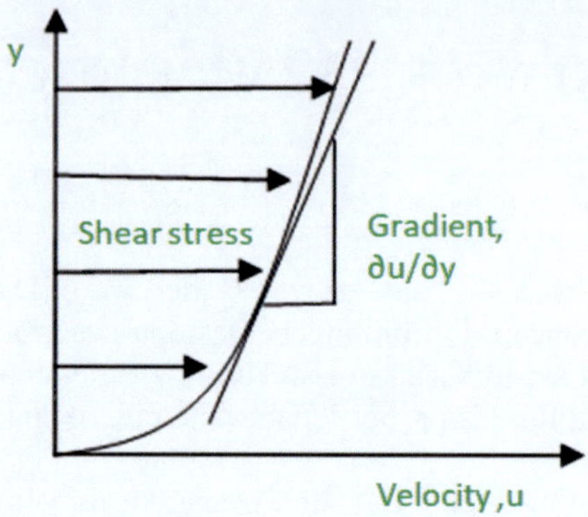

Applications of Newton's Law of Viscosity

Applications of Newton's Law of Viscosity include:

- **Fluid mechanics:** The Law of Viscosity is used in the study of fluids, including liquids and gases. It is used to describe the behavior of fluids in pipes, pumps, and other systems, as well as in the design of hydraulic machinery and other fluid-based technologies.
- **Petroleum engineering:** The Law of Viscosity is used in the study of petroleum fluids, including oil and natural gas. It is used to describe the flow of fluids through pipelines and other systems, as well as in the design of oil and gas drilling and production equipment.
- **Chemical engineering:** The Law of Viscosity is used in the study of chemical processes, including mixing and reaction processes. It is used to describe the behavior of fluids in chemical reactors, as well as in the design of chemical production processes.
- **Biomedical engineering:** The Law of Viscosity is used in the study of blood flow and other bodily fluids. It is used to describe the behavior of fluids in the circulatory system, as well as in the design of medical devices such as blood pumps and dialysis machines.

129. Newton's Laws of Motion

Newton's laws of motion are a set of three fundamental laws that describe the behavior of objects in motion. They were first presented by the English physicist and mathematician Sir Isaac Newton in his book "Philosophiæ Naturalis Principia Mathematica" in 1687. The three laws are:

- **Newton's first law:** Also known as the law of inertia, this law states that an object at rest will remain at rest, and an object in motion will remain in motion with a constant velocity, unless acted upon by an external force.

This law basically describes the tendency of objects to resist changes in their state of motion. Objects in motion tend to remain in motion, and objects at rest tend to remain at rest, unless acted upon by an external force. The law of inertia is particularly useful in predicting the behavior of objects in various situations, such as the motion of a ball after it has been hit by a bat, or the behavior of a car during a collision.

The first law of motion has a wide range of applications in our daily lives, from the functioning of simple machines to the operation of complex systems such as airplanes and rockets. Understanding the law of inertia is critical for engineers, scientists, and anyone working in fields related to mechanics and motion.

- **Newton's second law:** This law describes how the acceleration of an object is related to the net force acting on it and its mass. It states that the acceleration of an object is directly proportional to the net force acting on it, and inversely proportional to its mass. The mathematical expression of this law is

$$\mathbf{F = m\,a}$$

where F is the net force, m is the mass of the object, and a is its acceleration.

- **Newton's third law:** This law states that for every action, there is an equal and opposite reaction. When one object exerts a force on another object, the second object exerts an equal and opposite force on the first object.

These laws have profound implications for our understanding of the behavior of objects in the physical world, and they are used extensively in fields such as physics, engineering, and astronomy. They are also the foundation of classical mechanics, which is the study of the motion of objects under the influence of forces.

Newton's laws of motion in physics	
LAW #1	A body at rest will remain at rest, and a body in motion will remain in motion unless it is acted upon by an external force.
LAW #2	The force acting on an object is equal to the mass of that object times its acceleration, F = ma.
LAW #3	For every action, there is an equal and opposite reaction.

Applications of Newton's Laws of Motion

Newton's laws of motion are essential for understanding the behavior of objects in the physical world. Newton's laws of motion have numerous practical applications in different fields, including:

- **Transportation:** The principles of Newton's laws are used in the design and operation of various transportation systems, such as cars, airplanes, rockets, and spacecraft. The laws help engineers to determine the forces acting on the vehicle and how to control them for safe and efficient operation.
- **Sports:** Athletes and coaches use the principles of Newton's laws to improve performance in various sports, such as track and field, soccer, football, and baseball. The laws help to optimize the athlete's movements, such as running, jumping, throwing, and hitting.
- **Engineering:** Engineers use the principles of Newton's laws to design and develop various machines and structures, such as bridges, buildings, and machines. The laws help to ensure the safety and stability of the structures and the efficiency of the machines.
- **Astronomy:** Newton's laws are used in astronomy to understand the motion of planets, stars, and other celestial bodies. The laws help astronomers to calculate the orbits of planets, predict eclipses, and understand the motion of galaxies.

- **Medicine:** The principles of Newton's laws are used in medicine to understand the biomechanics of the human body, such as the movement of bones, joints, and muscles. The laws help to diagnose and treat injuries, design prosthetics, and improve rehabilitation.
- **Physics:** Newton's laws are fundamental principles in classical mechanics, which is the foundation of physics. They are used in many areas of physics, such as thermodynamics, fluid mechanics, and quantum mechanics.
- **Inertia:** Newton's First Law of Motion helps us to understand the concept of inertia, which is the tendency of an object to resist a change in its state of motion. This principle is applied in the design of safety equipment such as seat belts, airbags, and helmets, which are designed to protect passengers in case of a sudden change in the motion of the vehicle.
- **Friction:** Friction is an opposing force that resists motion between two surfaces. The first law of motion helps us to understand how friction works, and it is used in various applications such as brake pads in vehicles, anti-slip surfaces, and industrial machines.
- **Ballistics:** The laws of motion are applied in the field of ballistics, which involves the study of the motion of projectiles. It is used to determine the trajectory of a projectile, and to design weapons and ammunition.
- **Rockets and spacecraft:** Newton's Law of Motion are also used in the design and operation of spacecraft. The principle of action and reaction is used in rocket propulsion. The exhaust gases from the rocket engine are expelled out of the back of the rocket at high speed, creating a force in the opposite direction that propels the rocket forward.
- **Balloons:** When air is blown into a balloon, the balloon expands and the air inside exerts a force on the walls of the balloon. According to Newton's Third Law, the balloon will exert an equal and opposite force on the air inside, which causes the air to be pushed out of the balloon.
- **Swimming:** The principle of action and reaction is also used in swimming. When a swimmer pushes against the water with their hands and feet, the water exerts an equal and opposite force on the swimmer, propelling them forward.

130. NOETHER'S THEOREM

Noether's theorem is a fundamental theorem in physics that establishes a connection between symmetries and conservation laws. It was first proved by the mathematician Emmy Noether in 1915, and it has since become a cornerstone of modern theoretical physics.

The theorem states that every continuous symmetry of a physical system corresponds to a conserved quantity. In other words, if a physical system has a symmetry, such as translational symmetry, rotational symmetry, or time translation symmetry, then there is a corresponding conservation law, such as conservation of momentum, angular momentum, or energy, respectively.

This connection between symmetries and conservation laws is extremely powerful, and it is used in a wide range of areas in physics, including classical mechanics, electromagnetism, and quantum field theory. The theorem also has important implications for our understanding of the fundamental nature of space, time, and matter.

Emmy Noether

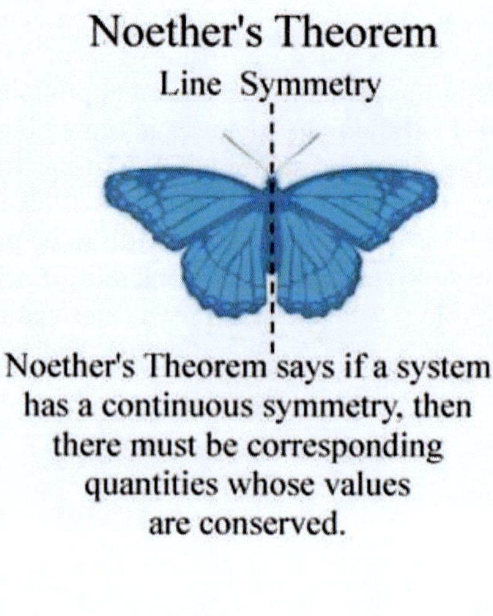

$$\frac{d}{dt}\left(\sum_a \frac{\delta L}{\delta \frac{dq_a}{dt}} \delta q_a\right) = 0$$

Applications of Noether's Theorem

Noether's theorem is a fundamental result in theoretical physics that relates symmetries in a physical system to conserved quantities. It has numerous applications in different branches of physics, including:

- **Classical mechanics:** Noether's theorem can be used to derive the conservation laws of energy, linear momentum, and angular momentum in classical mechanics from the symmetries of time translation, space translation, and rotation, respectively.
- **Quantum mechanics:** In quantum mechanics, Noether's theorem can be used to derive the conservation of probability, which is related to the symmetry of time translation.
- **Field theory:** Noether's theorem can be used to derive the conservation laws of energy, momentum, and angular momentum in field theory, as well as the conservation of electric charge and other conserved quantities.
- **General relativity:** In general relativity, Noether's theorem can be used to derive the conservation of energy and momentum, as well as the conservation of angular momentum in asymptotically flat space-times.
- **Particle physics:** Noether's theorem plays a crucial role in the Standard Model of particle physics, where it is used to derive the conservation of electric charge, color charge, and other conserved quantities.
- **Condensed matter physics:** Noether's theorem is used to study the properties of topological insulators and other topological phases of matter, where it relates the topological invariants of the system to the symmetries of the Hamiltonian.
- **Statistical mechanics:** Noether's theorem is used in statistical mechanics to derive the fluctuation-dissipation theorem, which relates the fluctuations and dissipation of a physical system to its symmetries.

Noether's theorem has far-reaching implications in many areas of physics, and it continues to be an active area of research.

131. OHM'S LAW

Ohm's Law is a fundamental law of electrical circuits that describes the relationship between the current flowing through a conductor, the voltage applied across it, and its resistance. It states that the current (I) flowing through a conductor is directly proportional to the voltage (V) applied across it and inversely proportional to its resistance (R). Mathematically, Ohm's Law can be expressed as:

$$V = I\,R$$

where V is the voltage applied across the conductor, I is the current flowing through it, and R is the resistance of the conductor.

This law is named after Georg Simon Ohm, a German physicist who first published it in 1827.

Georg Simon Ohm

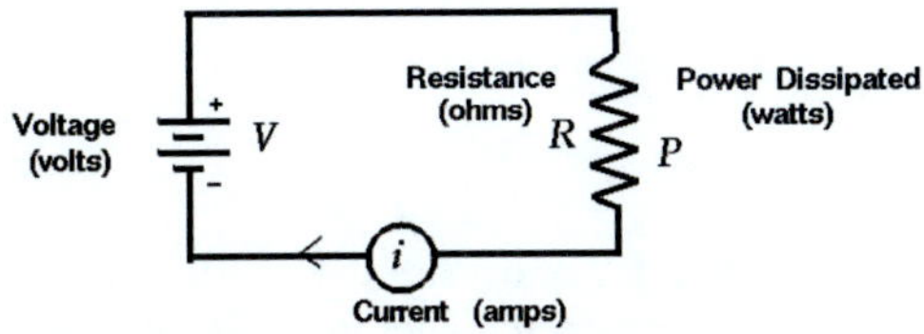

$$V = i\ R$$

$$P = i\ V = i^2 R$$

Applications of Ohm's Law

Ohm's Law is widely used in the design and analysis of electrical circuits, including power supplies, motors, and electronic devices. Some practical applications of Ohm's Law include:

♦ **Design of electrical circuits:** Ohm's Law is used to design and analyze electrical circuits, including voltage dividers, current limiting circuits, and power supplies. By applying the law, engineers can determine the appropriate values of resistors, capacitors, and other components to achieve the desired performance of the circuit.

♦ **Troubleshooting electrical circuits:** Ohm's Law is also used to troubleshoot electrical circuits and diagnose faults. By measuring the voltage and current in a circuit, and using the law to calculate the resistance, engineers can identify problems such as open circuits, short circuits, and faulty components.

♦ **Power calculations:** Ohm's Law is also used to calculate the power dissipated by a resistor in a circuit. The power (P) dissipated by a resistor can be calculated using the formula:

P = I V = I^2 R = V^2 / R

where I is the current flowing through the resistor, V is the voltage across it, and R is the resistance of the resistor. This formula is important in determining the power rating of resistors and other components in a circuit.

♦ **Electrical safety:** Ohm's Law is important in ensuring electrical safety in circuits and systems. By calculating the current flowing through a circuit, engineers can determine whether the circuit is safe for use and whether the components are capable of handling the current.

132. OSIPKOV–MERRITT MODEL

The Osipkov-Merritt model is a dynamical model used to describe the evolution of self-gravitating systems such as galaxies and star clusters. It was introduced by Russian astrophysicist Leonid Osipkov in 1979 and later refined by American astronomer David Merritt in 1985.

The model assumes that the system is spherically symmetric and that its stars move under the influence of both gravity and anisotropic velocity dispersion.

The Osipkov-Merritt model provides a way to calculate the density and velocity distribution of stars in the system as a function of radius, assuming a certain distribution function. The model is particularly useful for studying the dynamics of globular clusters, where the assumption of spherical symmetry is often a good approximation.

Applications of Osipkov–Merritt Model

The Osipkov-Merritt model is a theoretical model used to describe the distribution of stars in a globular cluster. It is based on the assumption that the stars in the cluster are in a state of dynamical equilibrium and move in a spherically symmetric potential. The model takes into account the effects of the stars' velocities and the distribution of their angular momenta.

- One of the key applications of the Osipkov-Merritt model is in understanding the structure and evolution of globular clusters. By fitting the model to observational data, astronomers can determine the distribution of stars within the cluster and make predictions about its evolution over time. The model can also be used to study the effects of various factors on the cluster, such as the presence of a central black hole or the effects of tidal forces from a nearby galaxy.
- Another application of the Osipkov-Merritt model is in understanding the dynamics of galactic nuclei. The model can be applied to study the distribution and motion of stars in the central regions of galaxies, where the effects of gravity are strongest. This can provide insights into the structure and evolution of galaxies, as well as the properties of supermassive black holes that are thought to reside at their centers.

133. Ostwald Dilution Law

Ostwald's dilution law states that the degree of dissociation of a weak electrolyte is inversely proportional to the square root of the concentration of the solution. This law was proposed by the German chemist Wilhelm Ostwald in 1887. This law is used to calculate the dissociation constant of weak electrolytes.

Wilhelm Ostwald

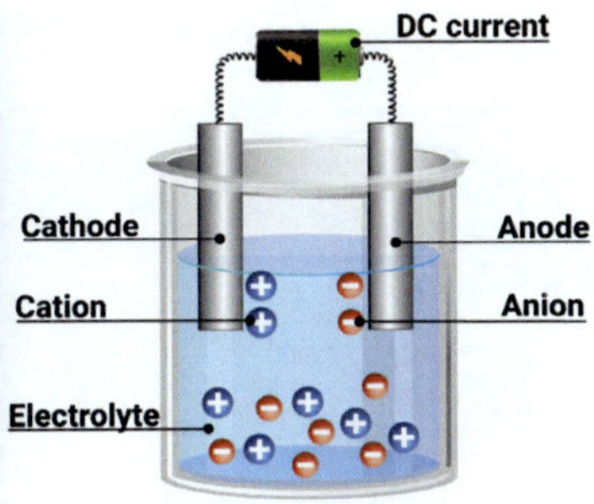

Applications of Ostwald Dilution Law

The Ostwald dilution law describes the relationship between the rate of reaction and the concentration of reactants in a solution. This law is often used in chemical kinetics to determine the rate constant of a reaction.

- One important application of the Ostwald dilution law is in the determination of dissociation constants (pKa) of weak acids and bases. The pKa value is a measure of the strength of an acid or base, and it represents the pH at which the acid or base is half dissociated. By measuring the pH of a solution at different dilutions, the pKa value can be calculated using the Ostwald dilution law.
- Another application of the Ostwald dilution law is in the study of reaction mechanisms. By monitoring the reaction rate at different concentrations, it is possible to determine the order of the reaction and the rate-determining step. This information is essential for the design and optimization of chemical processes.

134. Pascal's Law

Pascal's law is a fundamental principle in physics that describes the behavior of fluids (liquids and gases) under pressure. It states that a change in pressure applied to an enclosed fluid is transmitted uniformly throughout the fluid in all directions, without any loss of energy.

Mathematically, Pascal's law can be expressed as: "The pressure exerted anywhere in an enclosed fluid is transmitted equally in all directions throughout the fluid and acts with equal force on equal areas."

Blaise Pascal

(Credit:https://commons.wikimedia.org/wiki/File:Portrait_de_Blaise_Pascal_%28MARQ_999.3.1%29.jpg)

** PRINCIPLE **

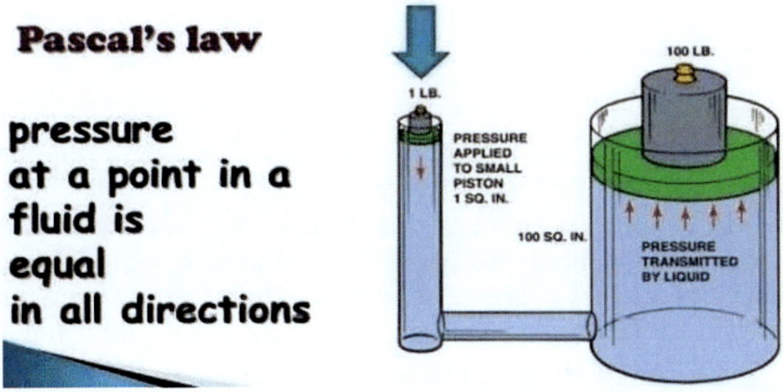

Applications of Pascal's Law

Pascal's law is a fundamental principle in physics that has many applications in various fields of science and engineering. It provides a powerful tool for understanding the behavior of fluids under pressure and has enabled the development of many important technologies, such as hydraulic systems and pressure gauges. Applications of Pascal's law include:

- **Hydraulic systems:** Pascal's law is the basis of hydraulic systems, which use fluids to transmit and amplify forces. Hydraulic systems are widely used in industrial machinery, construction equipment, and transportation systems.
- **Brakes:** Pascal's law is used in braking systems in vehicles, such as cars, trucks, and trains. By applying a small force to a pedal, the pressure is transmitted through a hydraulic system to the brake calipers, which apply a much larger force to the brake pads.
- **Pressure gauges:** Pascal's law is used in pressure gauges to measure the pressure of fluids. The gauge contains a fluid that is in contact with the fluid being measured, and the pressure of the fluid being measured is transmitted to the gauge through a tube or pipe.
- **Dams:** Pascal's law is used in the design of dams and reservoirs. The pressure of the water at the base of a dam is transmitted uniformly in all directions, allowing the dam to withstand the pressure and prevent water from flowing over the top.

135. PAULI EXCLUSION PRINCIPLE

The Pauli exclusion principle is a fundamental principle of quantum mechanics that states that no two identical fermions (particles with half-integer spin, such as electrons) can occupy the same quantum state simultaneously. This means that if one fermion is in a particular quantum state, no other fermion can be in that same quantum state.

The Pauli exclusion principle plays a crucial role in determining the electronic structure of atoms, molecules, and solids. It explains why atoms have discrete energy levels, and why chemical bonding occurs. It also explains many other properties of matter, such as the thermal and electrical conductivities of metals.

The principle was first formulated by the Austrian physicist Wolfgang Pauli in 1925, and is now considered one of the most important principles in all of physics.

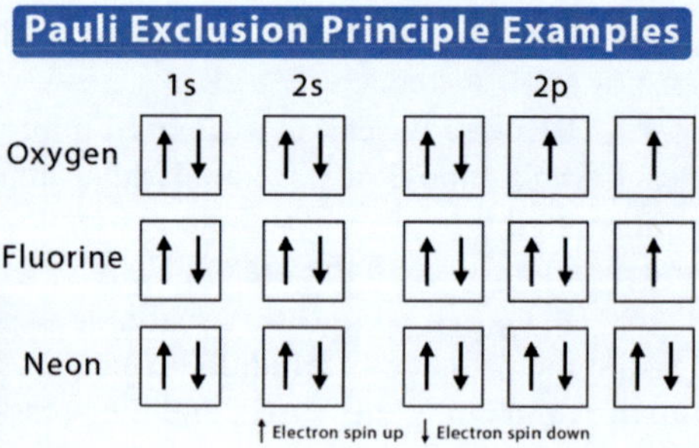

Wolfgang Pauli

(Credit: https://www.flickr.com/photos/193673378@N02/51376308875/)

Applications of Pauli Exclusion Principle

The Pauli exclusion principle is a fundamental principle in quantum mechanics that states that no two identical fermions (particles with half-integer spin) can occupy the same quantum state simultaneously. This principle has several important applications, including:

- **Electron configuration of atoms:** The Pauli exclusion principle explains why electrons in an atom occupy different energy levels and orbitals. This principle ensures that no two electrons in an atom can have the same set of quantum numbers (i.e., the same energy, angular momentum, and magnetic quantum numbers).
- **Chemical bonding:** The Pauli exclusion principle plays a crucial role in chemical bonding. When two atoms come together to form a molecule, their electrons interact and occupy different molecular orbitals. The Pauli exclusion principle ensures that the electrons in the same molecular orbital have different spin quantum numbers, which leads to the formation of chemical bonds.
- **Conductivity of metals:** The Pauli exclusion principle also explains why metals conduct electricity. In metals, the outermost electrons are free to move around and form a "sea" of electrons. The Pauli exclusion principle ensures that the electrons in this sea have different spin quantum numbers, which allows them to move through the metal without colliding with each other.
- **Neutron stars:** The Pauli exclusion principle is also responsible for the stability of neutron stars. In these extremely dense objects, the neutrons are packed so tightly together that they are subject to enormous pressures. The Pauli exclusion principle prevents the neutrons from collapsing into the same quantum state, which helps to support the neutron star against gravitational collapse.
- **Superconductivity:** The Pauli exclusion principle is also important in understanding superconductivity, a phenomenon in which certain materials can conduct electricity with zero resistance at low temperatures. The electrons in a superconductor form pairs called Cooper pairs, which behave like bosons (particles with integer spin). These Cooper pairs can occupy the same quantum state due to the Bose-Einstein statistics, which are not subject to the Pauli exclusion principle.

136. Planck's Equation

Planck's equation, also known as the Planck radiation law, describes the spectral density of electromagnetic radiation emitted by a black body at a given temperature. It was first formulated by the German physicist Max Planck in 1900 and is a fundamental law of physics. The equation is given as follows:

$$\mathbf{E = h\,\nu}$$

where,

E is the energy of a photon of electromagnetic radiation.

h is Planck's constant (6.626 x 10^-34 joule-seconds).

ν is the frequency of the radiation.

This equation relates the energy of a photon to its frequency. Planck's constant is a fundamental constant of nature that relates the energy of a photon to its frequency. The higher the frequency of the radiation, the higher the energy of the photon.

Planck's equation also forms the basis for quantum mechanics, which describes the behavior of particles at the atomic and subatomic level. It was a revolutionary breakthrough in the understanding of the behavior of electromagnetic radiation and has had numerous applications in physics, chemistry, and engineering.

Max Planck

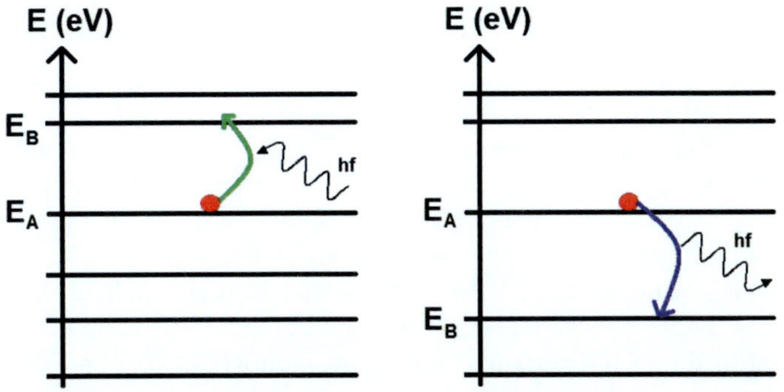

Absorption of a photon   Emission of a photon

$$\Delta E = |E_B - E_A| = h \times f = h \times \frac{c}{\lambda}$$

Applications of Planck's Equation

The law was first proposed by Max Planck in 1900, and later extended by Albert Einstein in 1905 to include the photoelectric effect. The law played a crucial role in the development of quantum mechanics. Planck's equation has numerous applications in a wide range of fields, including astrophysics, cosmology, spectroscopy, chemistry, and engineering. Some of the main applications are:

- **Understanding blackbody radiation:** Planck's equation is used to describe the spectrum of electromagnetic radiation emitted by a black body at a given temperature. A black body is an idealized object that absorbs all incident electromagnetic radiation and emits radiation at a characteristic temperature-dependent spectrum. Planck's equation is used to understand the radiation emitted by black bodies, which is important in fields such as astronomy and thermodynamics.
- **Semiconductor physics:** Planck's equation is used to understand the behavior of electrons in semiconductors. The energy of an electron is related to its frequency, and the behavior of electrons in semiconductors can be understood in terms of their energy levels.
- **Quantum mechanics:** Planck's equation is a fundamental equation of quantum mechanics, which describes the behavior of particles at the atomic and subatomic level. It is used to understand the behavior of photons, electrons, and other particles at the quantum level.

- **Spectroscopy:** Planck's equation is used in spectroscopy to analyze the absorption and emission spectra of molecules and atoms. By measuring the frequency of the absorbed or emitted radiation, the energy levels of the molecule or atom can be determined.
- **Lighting technology:** Planck's equation is used in lighting technology to determine the optimal color temperature for various applications. The color temperature of a light source is related to its spectrum, which can be described using Planck's equation.
- **Developing incandescent light bulbs:** The law helped develop incandescent light bulbs by understanding how different materials emit radiation at different temperatures, allowing scientists to design filaments that produce light at desired wavelengths.
- **Developing photovoltaic cells:** The law helps design photovoltaic cells by understanding the spectral distribution of radiation emitted by the sun and how it is absorbed by different materials.
- **Studying the cosmic microwave background radiation:** The Planck-Einstein law can be used to understand the cosmic microwave background radiation, which is the afterglow of the Big Bang.
- **Understanding the properties of materials:** The Planck-Einstein law can be used to understand the thermal properties of materials, such as their heat capacity and thermal conductivity.
- **Developing thermography:** The Planck-Einstein law has applications in thermography, which is the technique of using infrared radiation to measure the temperature of an object.

137. PLANCK'S LAW OF BLACK-BODY RADIATION

Planck's law of black-body radiation is a theoretical law that describes the spectral distribution of electromagnetic radiation emitted by a black body at a given temperature. It was formulated by German physicist Max Planck in 1900 and marked a major breakthrough in the development of quantum mechanics.

Planck's law states that the spectral radiance of a black body at a given temperature is proportional to the frequency of the radiation and to the fourth power of the absolute temperature. This relationship is expressed mathematically as:

$$B_\lambda(T) = \frac{2hc^2/\lambda^5}{e^{hc/\lambda kT} - 1}$$

where B is the spectral radiance of the black body at wavelength λ and temperature T, h is Planck's constant, c is the speed of light, k is Boltzmann's constant, and e is the base of the natural logarithm.

Planck's law has numerous applications in astrophysics, cosmology, and engineering, and has been tested and verified experimentally many times. It is considered one of the most important laws in physics and is a cornerstone of the modern understanding of quantum mechanics.

Max Planck

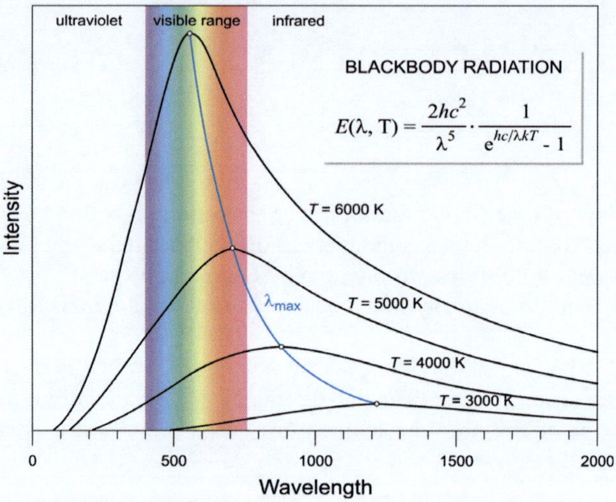

Applications of Planck's Law of Black-body Radiation

Planck's law of black-body radiation describes the spectral density of electromagnetic radiation emitted by a black body at a given temperature. Some of the applications of Planck's law of black-body radiation are:

♦ **Thermal radiation:** Planck's law is used to explain the thermal radiation emitted by a hot object. It is used in the study of thermodynamics and heat transfer.

♦ **Astrophysics:** Planck's law is used to study the temperature and composition of stars and galaxies. It is also used to calculate the temperature of the cosmic microwave background radiation, which is believed to be a remnant of the Big Bang.

♦ **Lighting design:** Planck's law is used to design light sources for different applications. For example, the color temperature of a light source can be adjusted by changing the temperature of the emitting material.

♦ **Solar cells:** Planck's law is used to optimize the efficiency of solar cells. By understanding the spectrum of sunlight, researchers can design solar cells that are optimized for maximum absorption of energy.

♦ **Infrared spectroscopy:** Planck's law is used in infrared spectroscopy to determine the chemical composition of a substance. By analyzing the absorption spectrum of a material, researchers can determine the chemical bonds present in the material.

138. POISEUILLE'S LAW

Poiseuille's law is a physical law that describes the steady flow of an incompressible and Newtonian fluid through a cylindrical pipe. It is named after the French physicist Jean-Louis Poiseuille, who derived the formula in the mid-19th century.

According to Poiseuille's law, the volume flow rate (Q) of a fluid through a cylindrical pipe of length (L), radius (r), and viscosity (η) is given by:

$$Q = (\pi\ r^4\ \Delta P) / (8\ \eta\ L)$$

where ΔP is the pressure difference between the ends of the pipe. The law states that the volume flow rate is directly proportional to the fourth power of the radius of the pipe and the pressure difference, and inversely proportional to the viscosity and the length of the pipe.

Poiseuille's law is important in the study of fluid mechanics and is widely used in engineering applications, such as the design of pipelines, blood flow in vessels, and flow in microfluidic devices.

Jean-Louis Poiseuille

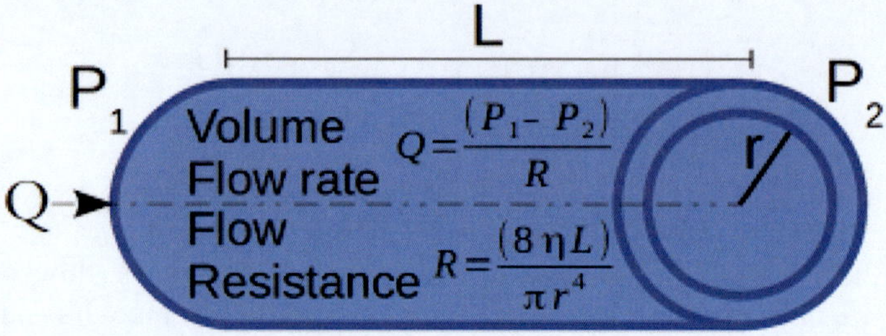

Applications of Poiseuille's Law

Poiseuille's law describes the flow of a viscous fluid through a cylindrical pipe or tube. It has several applications in the fields of physics, engineering, and medicine:

- **Blood flow:** Poiseuille's law is used to understand the flow of blood in the circulatory system. It helps to predict the resistance to blood flow in the arteries and veins, which is related to various diseases such as hypertension.
- **Fluid dynamics:** Poiseuille's law is used to describe fluid flow in pipes and tubes, and is applied in the design of piping systems in chemical plants, water supply networks, and HVAC systems.
- **Microfluidics:** Poiseuille's law is also used in microfluidics, which is the study and manipulation of fluids at a microscopic scale. It helps to understand the flow of fluids through microchannels and microfluidic devices, which have applications in lab-on-a-chip technology, biosensors, and drug delivery systems.
- **Industrial applications:** Poiseuille's law is used in various industrial applications such as inkjet printing, fuel injection systems, and lubrication systems. It helps to predict the pressure drop and flow rate of fluids through pipes and channels.
- **Geophysics:** Poiseuille's law is used to study the flow of fluids through porous media such as rocks and soils, which has applications in hydrology, petroleum engineering, and geophysics. It helps to understand the flow of groundwater, oil, and gas in subsurface reservoirs.

139. PYTHAGORAS THEOREM

The Pythagorean theorem is a fundamental principle in mathematics that relates to the sides of a right-angled triangle. It states that in a right-angled triangle, the square of the length of the hypotenuse (the side opposite the right angle) is equal to the sum of the squares of the lengths of the other two sides.

In mathematical terms, if we have a right-angled triangle with side lengths a, b, and c (where c is the hypotenuse), the Pythagorean theorem can be expressed as:

$$a^2 + b^2 = c^2$$

This theorem is attributed to the ancient Greek mathematician Pythagoras, who lived around 500 BCE. However, it is important to note that the Pythagorean theorem was known and used in various cultures even before Pythagoras.

The Pythagorean theorem has numerous applications in mathematics, physics, engineering, and other fields. It is used to solve problems involving right-angled triangles, such as finding unknown side lengths or determining whether a triangle is right-angled based on its side lengths.

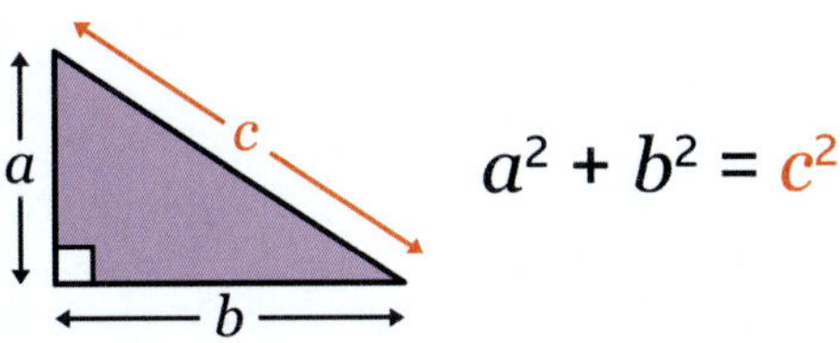

Applications of Pythagoras Theorem

The Pythagorean theorem has a wide range of applications in various fields. Its versatility and practicality make it an essential mathematical principle used in numerous real-world scenarios. Here are some notable applications:

Geometry and Trigonometry: The Pythagorean theorem is extensively used in geometry and trigonometry to solve problems involving right-angled triangles. It allows us to find unknown side lengths, determine angles, and prove geometric relationships.

Construction and Architecture: Builders and architects use the Pythagorean theorem to ensure accurate and precise measurements when constructing right angles. It is essential for maintaining square corners and perpendicular lines in buildings.

Navigation and GPS: The Pythagorean theorem plays a crucial role in navigation and GPS systems. It is used to calculate distances and positions based on the coordinates of two points on a map or globe.

Engineering: Engineers use the Pythagorean theorem for various applications, including structural analysis, designing electrical circuits, surveying land, and calculating forces and vectors.

Computer Graphics: In computer graphics and 3D modeling, the Pythagorean theorem is utilized to determine the distance between two points in a 3D space. It is essential for rendering realistic images, simulating physics, and creating virtual environments.

Optics: The Pythagorean theorem is employed in optics to analyze the path of light rays. It helps in determining the focal length of lenses, calculating image distances, and understanding the behavior of light in optical systems.

Physics: The Pythagorean theorem is used in physics for vector analysis and solving problems involving forces and motion. It is utilized in calculating components of vectors, determining resultant forces, and analyzing projectile motion.

Surveying and Cartography: Surveyors and cartographers rely on the Pythagorean theorem to measure distances, establish boundaries, and create accurate maps.

Everyday Life: The Pythagorean theorem finds applications in everyday situations, such as calculating the diagonal length of a TV or computer screen, determining the length of a ladder required to reach a specific height, or estimating the distance traveled in a zigzag path.

140. Radioactive Decay Law

The radioactive decay law is a fundamental principle in physics that describes the process by which radioactive isotopes spontaneously decay into more stable isotopes over time. It states that the rate of radioactive decay of a particular isotope is proportional to the number of radioactive atoms present at any given time.

The decay rate is typically measured by the half-life of a radioactive sample, which is the time it takes for half of the radioactive atoms in the sample to decay. The half-life is a characteristic property of each radioactive element and can range from fractions of a second to billions of years.

Mathematically, the law can be expressed as:

$$N(t) = N_0 e^{(-\lambda t)}$$

where N(t) is the number of radioactive atoms remaining at time t, N_0 is the initial number of radioactive atoms, λ is the decay constant, and e is the mathematical constant known as Euler's number.

The half-life of a radioactive isotope is the amount of time it takes for half of the initial number of radioactive atoms to decay. It can be calculated using the decay constant as:

$$t_{1/2} = \ln(2) / \lambda$$

where ln (2) is the natural logarithm of 2, which is approximately 0.693. The half-life is a useful parameter for understanding the rate of radioactive decay and the behavior of radioactive isotopes over time.

Radioactivity was discovered in 1896 by scientists Henri Becquerel and Marie Curie. Marie Curie, a Polish-born physicist and chemist, made significant contributions to the study of radioactivity and its laws. While she didn't directly discover the radioactive decay law, her work was instrumental in its development and understanding. Marie Curie, along with her husband Pierre Curie, conducted extensive research on radioactivity. In 1898, they discovered two new radioactive elements, which they named polonium and radium. Marie Curie conducted meticulous measurements of the radioactive decay rates of various radioactive elements. She devised techniques to accurately measure the emitted radiation and studied the decay properties of radium and other elements. She worked on refining techniques for measuring radioactivity, including the development of the Curie unit (named in honor of her and Pierre Curie), which quantified the intensity of radioactive emissions. This unit provided a standardized measure for comparing and characterizing different radioactive substances.

Ernest Rutherford

Frederick Soddy

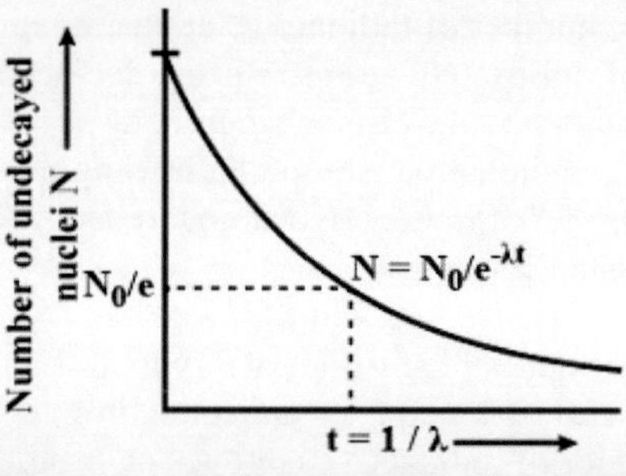

Henri Becquerel

Marie Curie

Applications of Radioactive Decay Law

The radioactive decay law has many important applications, including:

- **Radiometric dating:** The radioactive decay law is used in radiometric dating to determine the age of rocks, minerals, and other materials. By measuring the decay rate of a radioactive element and comparing it to the amount of the element remaining in a sample, scientists can determine how long it has been since the sample was last heated or otherwise altered.
- **Nuclear energy:** The radioactive decay law is fundamental to the production of nuclear energy. Nuclear reactors use the decay of radioactive isotopes, such as uranium-235, to produce heat that is used to generate electricity.
- **Medical imaging:** The radioactive decay law is used in medical imaging techniques, such as positron emission tomography (PET), to detect the presence of radioactive isotopes in the body. These isotopes are often used to track the uptake and distribution of drugs or to identify diseased tissues.
- **Environmental monitoring:** The radioactive decay law is used in environmental monitoring to measure the levels of radioactive contamination in air, water, soil, and other environmental samples. By measuring the decay rate of specific isotopes, scientists can determine the amount of radioactive material present in a sample and assess the potential risk to human health and the environment.

141. RAMAN SCATTERING

Raman scattering is a phenomenon in which a photon interacts with a molecule, causing the molecule to undergo a change in its vibrational energy state. This change in energy state results in the scattered photon having a different energy and frequency than the incident photon.

Raman scattering was discovered by the Indian Physicist Sir C. V. Raman in 1928 and it is named after him. Raman scattering is used in a variety of scientific and industrial applications, including in the study of molecular vibrations and in the detection of chemicals and biological molecules.

There are two types of Raman scattering: Stokes and anti-Stokes. Stokes Raman scattering occurs when a photon loses energy and is scattered at a lower frequency, while anti-Stokes Raman scattering occurs when a photon gains energy and is scattered at a higher frequency.

Raman scattering is a powerful tool for the identification and characterization of molecules. It is widely used in chemistry, biology, and materials science to study the structure and properties of molecules and materials.

C. V. Raman

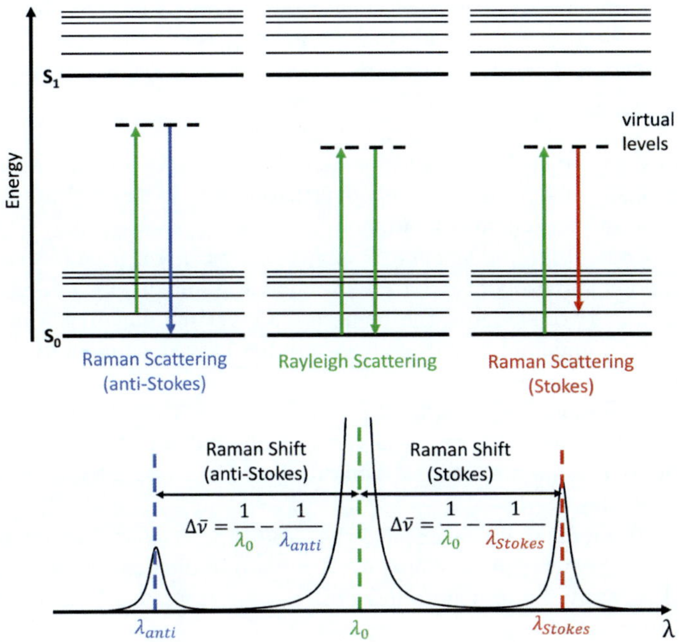

Applications of Raman Scattering

Raman scattering is a powerful technique with numerous applications in various fields, including chemistry, materials science, biomedical research, nanotechnology, and environmental science. It provides a non-destructive and non-invasive way to study the properties and composition of different types of samples, making it a valuable tool for scientific research and development. Some of the major applications of Raman scattering are:

- **Chemistry:** Raman spectroscopy is a powerful tool for studying the vibrational and rotational modes of molecules. It is used to identify and characterize chemical compounds, monitor chemical reactions, and study the molecular structure and conformation of complex molecules of a wide range of materials, including solids, liquids, and gases.

- **Materials science:** Raman spectroscopy is used to study the structure and properties of various materials, including semiconductors, metals, and polymers. It provides information about the crystal structure, defects, and impurities in materials, and can be used to study the phase transitions and mechanical properties of materials.

♦ **Biomedical research:** Raman spectroscopy is used in biomedical research to study the chemical composition of tissues and cells. It can be used to detect and diagnose diseases, monitor drug delivery, and study the interaction between drugs and biomolecules.

♦ **Nanotechnology:** Raman spectroscopy is used to study the properties of nanoparticles and nanomaterials. It provides information about the size, shape, and structure of nanoparticles, and can be used to monitor their synthesis and functionalization.

♦ **Environmental science:** Raman spectroscopy is used to study the composition and properties of environmental samples, including soils, sediments, and pollutants. It can be used to detect and quantify pollutants, monitor environmental changes, and study the biogeochemical cycling of nutrients and pollutants.

♦ **Pharmaceutical industry:** Raman scattering is used in the pharmaceutical industry for the analysis of drugs and drug delivery systems. It can be used to monitor drug dissolution, stability, and interactions with other molecules.

♦ **Biomedical research:** Raman scattering is used in biomedical research for the detection and characterization of biological samples. It can be used to study the structure and function of proteins, DNA, and other biological molecules.

♦ **Forensic science:** Raman scattering is used in forensic science for the analysis of trace evidence, including drugs, explosives, and fibers. It can be used to identify and characterize the composition of materials found at crime scenes.

142. Raoult's Law

Raoult's law states that the vapor pressure of a solution is directly proportional to the mole fraction of the solvent in the solution. In other words, as the amount of solvent in the solution increases, the vapor pressure of the solution also increases. This law applies to ideal solutions, where the interactions between the solvent and solute molecules are similar to those between the solvent molecules themselves.

François-Marie Raoult
(Credit: https://wellcomecollection.org/works/etds7jma)

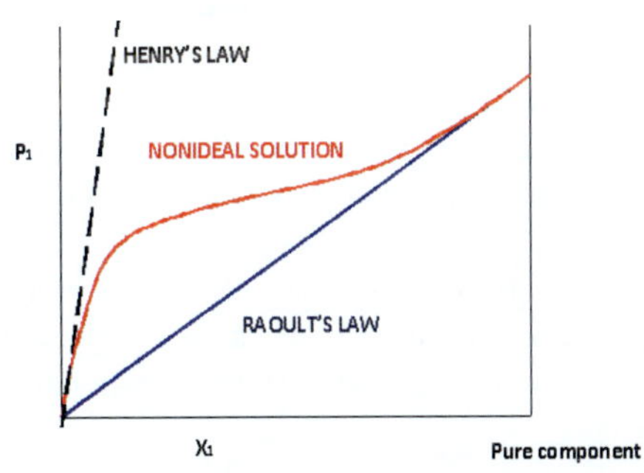

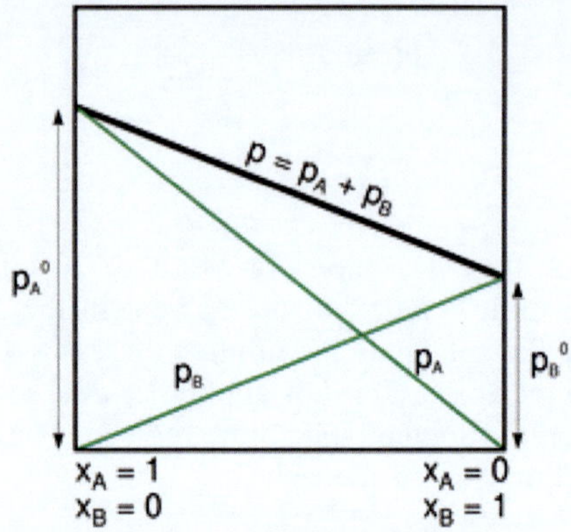

Applications of Raoult's Law

Raoult's law is a thermodynamic law that states that the vapor pressure of an ideal solution is proportional to the mole fraction of each component in the solution. Here are some applications of Raoult's law:

- **Vapor-liquid equilibria:** Raoult's law is used to calculate the vapor pressure of a solution at a given temperature, given the mole fraction of each component. This is useful for predicting the behavior of mixtures of liquids, such as in distillation.
- **Chemical engineering:** Raoult's law is used in the design and optimization of distillation columns, which are commonly used in the chemical industry to separate and purify different components in a mixture.
- **Electrochemistry:** Raoult's law can be used to calculate the activity coefficients of electrolytes in solution. This is important in electrochemistry, where the activity coefficients of the different ions in solution can affect the behavior of the electrochemical cell.
- **Atmospheric science:** Raoult's law is used to model the behavior of the atmosphere, particularly in the formation of clouds and precipitation. Cloud droplets are formed when water vapor condenses onto particles in the air, and the vapor pressure of the water in the cloud is governed by Raoult's law.
- **Pharmaceutical industry:** Raoult's law is used in the pharmaceutical industry to predict the solubility of drugs in different solvents. This is important for determining the appropriate formulation and delivery method for different drugs.

143. SAHA IONIZATION EQUATION

The Saha ionization equation is a mathematical equation that describes the degree of ionization of a gas at a certain temperature and pressure. It was developed by Indian astrophysicist Meghnad Saha in 1920.

The equation is particularly useful for studying the behavior of ionized gases, such as those found in stars. It relates the degree of ionization of a gas to the temperature, pressure, and ionization energy of the gas.

The Saha ionization equation is given by:

$$\frac{N_{II}}{N_I} = \frac{2Z_{II}}{n_e Z_I}\left(\frac{2\pi m_e kT}{h^2}\right)^{3/2} e^{-\chi/kT}$$

where, N_{II} and N_I are the number densities of the ionized and neutral species, respectively

Z_{II} and Z_I are the statistical weights of the ionized and neutral species, respectively

m_e is the mass of an electron

k is Boltzmann's constant

T is the temperature of the gas

h is Planck's constant

χ is the ionization energy of the species from level I to level II.

The equation shows that as the temperature increases, the degree of ionization of a gas also increases. This relationship is important for understanding the spectra of stars, as the degree of ionization affects the absorption and emission lines in a star's spectrum.

Meghnad Saha

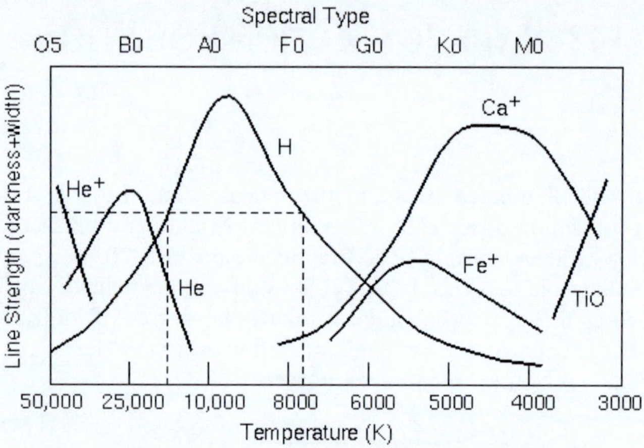

Applications of Saha Ionization Equation

The Saha ionization equation is a formula that relates the degree of ionization of a gas to the temperature and pressure of the gas, as well as the ionization energy of the atoms and the density of the gas. It is commonly used in astrophysics to study the ionization state of gases in stars and other astronomical objects.

- The Saha equation can be used to determine the relative abundances of different ionization stages of elements in stars, which can provide information about the temperature, density, and composition of the star. It can also be used to study the ionization of gases in the interstellar medium, which can give insight into the physical conditions and chemical evolution of galaxies.
- In addition, the Saha equation has applications in plasma physics and fusion research, where it is used to study the ionization state of plasmas and the conditions required for fusion reactions to occur.

144. SCHRÖDINGER EQUATION

The Schrödinger equation is a fundamental equation in quantum mechanics that describes how the wave function of a physical system changes over time. It is named after Austrian physicist Erwin Schrödinger, who first formulated it in 1925.

The Schrödinger equation is a partial differential equation that is used to describe the wave-like behavior of particles in quantum mechanics. It relates the time derivative of the wave function to its second spatial derivative and the potential energy of the system.

The equation is written as:

$$I \hbar \, \partial\psi/\partial t = H \psi$$

where i is the imaginary unit, $\hbar$ is the reduced Planck constant, ψ is the wave function of the system, t is time, and H is the Hamiltonian operator, which represents the total energy of the system.

It has been used to explain phenomena such as the double-slit experiment, quantum tunneling, and the behavior of atoms and molecules in chemical reactions.

Erwin Schrödinger

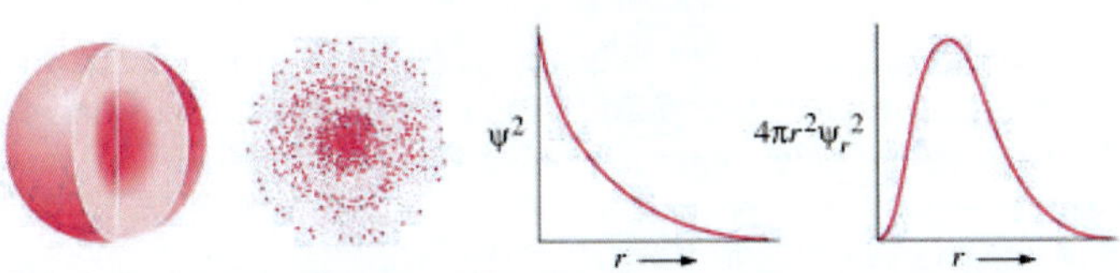

(a) (b) (c) (d)

Applications of Schrödinger Equation

The Schrödinger equation is a fundamental equation in quantum mechanics, which describes how the wave function of a physical system evolves over time. Its applications are broad and span across many areas of science and technology.

- **Atomic and molecular physics:** The Schrödinger equation provides a mathematical framework to describe the quantum states of atoms and molecules. It is used to calculate the energy levels and spectra of atoms and molecules, and to study the behavior of electrons in chemical reactions.
- **Solid state physics:** The Schrödinger equation is used to describe the electronic properties of materials, such as their conductivity, magnetism, and optical properties. It is used to model the behavior of electrons in semiconductors, superconductors, and other materials.
- **Quantum computing:** The Schrödinger equation provides the basis for the development of quantum algorithms and quantum computers. The evolution of the wave function can be used to perform quantum computations, which have the potential to solve certain problems much faster than classical computers.
- **Quantum field theory:** The Schrödinger equation is used to describe the behavior of quantum fields, which are fundamental to our understanding of particle physics. It is used to study the interactions between particles and the forces that govern them.
- **Astrophysics:** The Schrödinger equation is used to study the behavior of matter in extreme conditions, such as in the interiors of stars and black holes. It is used to calculate the properties of white dwarfs, neutron stars, and other exotic objects.
- **Chemistry:** The Schrödinger equation is used to study the structure and properties of molecules, and to design new drugs and materials. It is used to calculate the electronic structure of molecules, and to simulate chemical reactions.
- **Materials science:** The Schrödinger equation is used to study the properties of materials, such as their electronic and optical properties. It is used to design new materials with specific properties, such as high strength or conductivity.
- **Nanotechnology:** The Schrödinger equation is used to study the behavior of electrons and other particles in nanoscale materials and devices. It is used to design new nanomaterials and devices, such as nanoscale sensors and transistors.

145. SEEBECK EFFECT

The Seebeck effect is the thermoelectric phenomenon in which a temperature difference between two dissimilar electrically conducting materials produces a voltage difference between them. When one end of a thermocouple is heated, electrons within the material move towards the cooler end, generating a voltage across the two ends. This voltage is directly proportional to the temperature difference between the two ends of the thermocouple. The Seebeck effect is used in thermocouples, which are widely used in temperature measurement and control applications.

Thomas Johann Seebeck

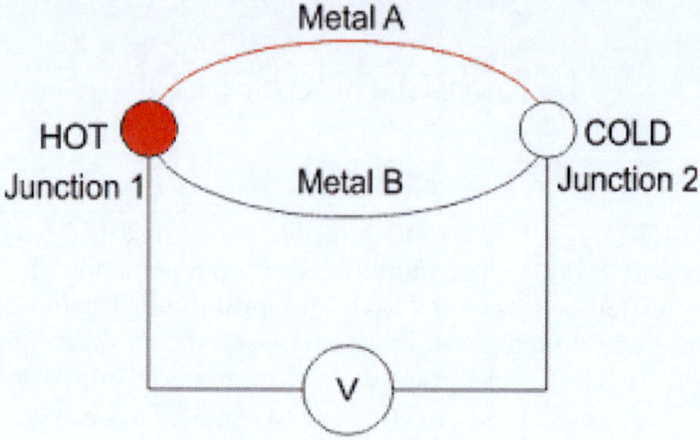

Applications of Seebeck Effect

The Seebeck effect refers to the phenomenon of generating a voltage across a material when there is a temperature difference across it. Seebeck effect has a wide range of applications in various fields, including energy conversion, temperature measurement, and thermal imaging.

- **Thermoelectric generators:** The Seebeck effect is the underlying principle behind thermoelectric generators, which convert heat directly into electrical energy. These generators are used in a variety of applications, including power generation in remote locations and waste heat recovery in industrial processes.
- **Temperature sensors:** The Seebeck effect can be used to measure temperature differences in a material. This is the principle behind thermocouples, which are used to measure temperature in a wide variety of industrial and scientific applications.
- **Heat pumps:** The Seebeck effect is used in some types of heat pumps, which can be used to heat or cool a space by transferring heat from one location to another.
- **Thermal imaging:** The Seebeck effect is used in some thermal imaging cameras, which can detect and display temperature differences in a scene.
- **Radioisotope thermoelectric generators:** The Seebeck effect is also used in radioisotope thermoelectric generators (RTGs), which are used to power spacecraft and other remote applications. RTGs use the heat generated by the radioactive decay of isotopes to generate electrical power using thermoelectric materials.

146. SÉRSIC'S LAW

Sérsic's law is a mathematical formula that describes the surface brightness profile of galaxies. It is named after the Argentine astronomer José Luis Sérsic, who introduced the law in 1968. The law states that the intensity of light from a galaxy decreases with distance from its center according to an exponential function, with an index that is a free parameter. This index is known as the Sérsic index, and it characterizes the shape of the galaxy's surface brightness profile. The Sérsic law is widely used in the field of extragalactic astronomy to model the brightness distribution of galaxies, and it has been shown to provide an excellent fit to observations.

Sérsic's law is a mathematical formula used to describe the surface brightness of galaxies. It is given by:

$$\log I(R) \propto R^{1/n}$$

where,

R is distance from brightest point.

I(R) is brightness (intensity) at R.

n is Sérsic index.

Most galaxies are fit by Sérsic profiles with indices in the range $1/2 < n < 10$. The best-fit value of n correlates with galaxy size and luminosity, such that bigger and brighter galaxies tend to be fit with larger n.

José Luis Sérsic

(Credit: https://academictree.org/physics/peopleinfo.php?pid=783832)

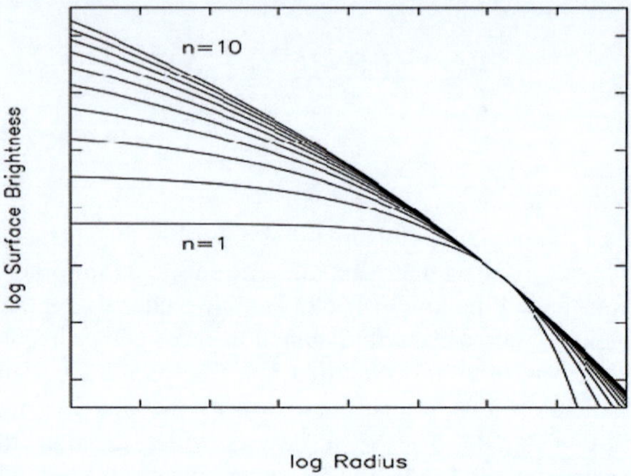

Applications of Sérsic's Law

Sérsic's law has found several applications in the field of astronomy and astrophysics, such as:

- **Galaxy morphology:** Sérsic's law is used to describe the light distribution in galaxies, which can be used to classify them based on their morphology. For example, galaxies with a low value of \$n\$ are typically disk-like, while those with a high value of \$n\$ are more centrally concentrated and have a more spherical shape.
- **Stellar populations:** Sérsic's law can be used to model the distribution of stars in a galaxy, and to estimate the total mass and luminosity of the galaxy. By fitting the observed surface brightness profile to the Sérsic law, astronomers can also infer the age and metallicity of the stellar populations.
- **Galaxy evolution:** By comparing the Sérsic parameters of galaxies at different redshifts, astronomers can study how the properties of galaxies have evolved over cosmic time. For example, the concentration parameter \$n\$ is found to increase with redshift, indicating that galaxies were more compact in the past.
- **Gravitational lensing:** Sérsic's law can be used to model the surface brightness distribution of galaxies that act as gravitational lenses. By comparing the observed image of a lensed galaxy to the predicted image from the Sérsic model, astronomers can infer the mass distribution of the lensing galaxy and study the properties of dark matter.

147. SOKOLOV–TERNOV EFFECT

The Sokolov-Ternov effect refers to the phenomenon of the spin rotation of a relativistic particle in a magnetic field, which occurs due to the interaction between the particle's magnetic moment and the external magnetic field. The effect is named after Arseny Alexandrovich Sokolov and Igor Mikhailovich Ternov, who first described it in 1968.

The Sokolov-Ternov effect is important in the study of high-energy physics and is used in particle accelerators to measure the spin of particles. It has also been observed in astrophysical phenomena, such as the radiation emitted by pulsars.

Arseny Sokolov **Igor Ternov**

(Credit: https://smena-online.ru/sites/default/files/06_-_1950.pdf)
(Credit: https://www.phys.msu.ru/rus/about/sovphys/ISSUES-2011/07(91)-2011/10280/)

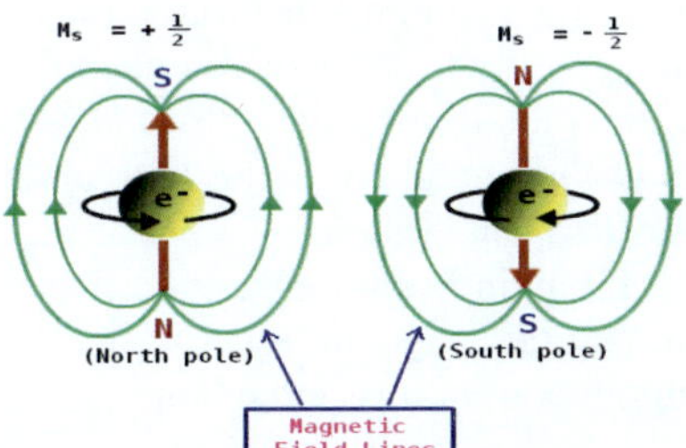

Applications of Sokolov–Ternov Effect

The Sokolov-Ternov effect, also known as the quantum synchrotron radiation, has a wide range of applications in accelerator physics, particle physics, astrophysics, and quantum science. It enables precise beam diagnostics, aids in understanding the behavior of charged particles in strong fields, and provides valuable insights into fundamental physical processes. Here are some notable applications:

- **Particle Accelerators:** The Sokolov-Ternov effect plays a crucial role in the design and operation of particle accelerators, such as synchrotrons and storage rings. By understanding and controlling the radiation emitted by relativistic electrons, scientists can optimize the performance of particle accelerators, minimize energy loss, and improve beam stability.
- **Beam diagnostics:** The emitted synchrotron radiation can be used as a diagnostic tool to measure the properties of charged particle beams. By analyzing the spectral and angular distribution of the radiation, scientists can determine the energy, momentum, and other characteristics of the electron beam, providing valuable information for beam tuning and optimization.
- **Polarimetry:** The polarization of synchrotron radiation generated by the Sokolov-Ternov effect carries information about the spin orientation of the emitting electrons. This property is exploited in polarimeters, devices used to measure the polarization state of the electron beam. Polarimetry is crucial in high-energy physics experiments and studies involving polarized electron beams.
- **Electrodynamics in strong fields:** The Sokolov-Ternov effect offers insights into the behavior of charged particles in strong electromagnetic fields. By studying the radiation spectrum and polarization, scientists can probe the quantum electrodynamics (QED) effects and test theoretical predictions in extreme conditions, such as in the presence of strong magnetic fields.
- **Astrophysics:** The quantum synchrotron radiation has implications in astrophysics, particularly in the study of compact objects like pulsars and black holes. The emitted radiation carries information about the properties of these objects, such as their magnetic fields and relativistic motion.
- **Quantum optics:** The Sokolov-Ternov effect has connections to quantum optics and quantum information science. The coherence properties of the emitted radiation are of interest for applications in quantum communication, quantum computing, and quantum metrology.

148. Sommerfeld–Kossel Displacement Law

The Sommerfeld–Kossel displacement law states that the first spark (singly ionized) spectrum of an element is similar in all details to the arc (neutral) spectrum of the element preceding it in the periodic table. Likewise, the second (doubly ionized) spark spectrum of an element is similar in all details to the first (singly ionized) spark spectrum of the element preceding it, or to the arc (neutral) spectrum of the element with atomic number two less, and so forth.

Hence, the spectra of C I (neutral carbon), N II (singly ionized nitrogen), and O III (doubly ionized oxygen) atoms are similar, apart from shifts of the spectra to shorter wavelengths. C I, N II, and O III all have the same number of electrons, six, and the same ground-state electron configuration:

$$1s^2\ 2s^2\ 2p^2\ 3p_0$$

The law was discovered by and named after Arnold Sommerfeld and Walther Kossel, who set it forth in a paper in 1919.

It's worth noting that as our understanding of atomic structure and spectroscopy advanced, the periodic table based on elemental properties and spectral patterns was refined and replaced by the modern periodic table, which is based on the atomic number and electron configurations of elements. While the first spark spectrum of a singly ionized element and the arc spectrum of the neutral element preceding it can share similarities, there can still be distinct differences due to the change in electronic structure upon ionization.

Arnold Sommerfeld **Walther Kossel**

(Credit: https://www.flickr.com/photos/193673378@N02/51375535933/)

(Credit: https://academictree.org/physics/peopleinfo.php?pid=735496)

149. Stefan-Boltzmann Law

The Stefan-Boltzmann law is a fundamental law of physics that describes the relationship between the temperature of an object and the amount of radiation it emits. It states that the total energy radiated per unit surface area of an object per unit time is proportional to the fourth power of its temperature. The law is named after the Austrian physicist Josef Stefan and the German physicist Ludwig Boltzmann.

The Stefan-Boltzmann law is expressed mathematically as:

E = σ T^4

where E is the total energy radiated per unit surface area per unit time, T is the temperature of the object in kelvins, and σ is the Stefan-Boltzmann constant, which has a value of approximately 5.67 x 10^-8 watts per square meter per kelvin to the fourth power.

Josef Stefan

Ludwig Boltzmann

$$\frac{P}{A} = \sigma T^4 \; j/m^2 s$$ Stefan-Boltzmann Law

$$\sigma = 5.6703x10^{-8}\, watt / m^2 K^4$$

Applications of Stefan-Boltzmann Law

The Stefan-Boltzmann Law has numerous applications across various fields of science and engineering. The law applies to all objects that radiate energy in the form of electromagnetic waves, including stars, planets, and even human bodies. It is particularly important in the study of astrophysics, where it is used to calculate the temperatures and luminosities of stars and other celestial objects. The Stefan-Boltzmann law also has practical applications in engineering and materials science.

- **Astrophysics:** The Stefan-Boltzmann Law is used extensively in astrophysics to determine the temperature, luminosity, and size of stars. By measuring the total energy emitted by a star, astronomers can estimate its temperature and the rate at which it is losing energy.
- **Climate science:** The Stefan-Boltzmann Law is also used in climate science to calculate the amount of energy radiated by the Earth and the radiative properties of greenhouse gases. This information is critical for understanding the Earth's energy balance and the impact of human activities on the climate.
- **Materials science:** The Stefan-Boltzmann Law is used in materials science to understand the thermal properties of materials. It is used, for example, to calculate the rate of heat transfer between two surfaces, and to design materials for use in high-temperature environments.
- **Industrial applications:** The Stefan-Boltzmann Law is used in various industrial applications such as in designing heating systems for buildings, designing ovens and kilns, and in developing high-temperature materials for use in aircraft engines and power plants.
- **Astronomy:** The Stefan-Boltzmann Law is also used in astronomy to estimate the temperature of celestial objects such as planets, moons, and asteroids.

150. STOKES' LAW

Stokes' law is a physical law that describes the behavior of a small spherical particle in a viscous fluid. It states that the drag force experienced by a small spherical particle moving through a viscous fluid is proportional to the velocity of the particle, the radius of the particle, the viscosity of the fluid, and the acceleration due to gravity.

The law was first derived by the British Physicist Sir George Gabriel Stokes in 1851. It is important in a number of areas, including fluid mechanics, sedimentation, and the study of the motion of small particles in the atmosphere and in water.

Stokes' law is given by the equation:

$$\mathbf{F = 6\pi\, r\, \eta\, v}$$

where F is the drag force on the particle, r is the radius of the particle, η is the viscosity of the fluid, and v is the velocity of the particle. The constant 6π is known as the Stokes drag coefficient.

Stokes' law is valid only for small particles moving slowly through a viscous fluid. For larger particles or higher velocities, other factors, such as turbulence, must be taken into account.

George Gabriel Stokes

Stokes' Law

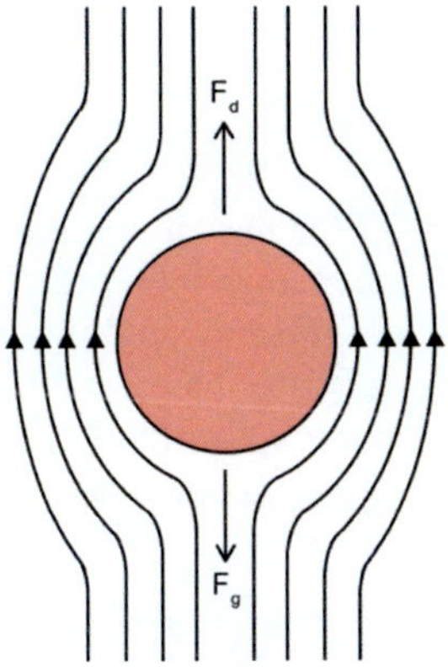

$$F_d = 6\pi\eta r v$$

η : Viscosity of the fluid

r : Radius of the particle

v : Velocity of the particle

F_d : Drag force

F_g : Gravitational force

Applications of Stokes' Law

Stokes's law is a fundamental principle in fluid mechanics and has numerous applications in various fields of science and engineering, some of which include:

- **Determination of viscosity:** Stokes's law provides a method for determining the viscosity of a fluid by measuring the terminal velocity of a small sphere moving through the fluid under the influence of gravity.
- **Sedimentation:** Stokes's law explains the settling behavior of particles in a fluid under the influence of gravity. This is useful in sedimentation processes such as water treatment, wastewater treatment, and mineral processing.
- **Rheology:** The viscosity of fluids plays an important role in rheology, which is the study of the flow of fluids. Stokes's law is used to model laminar flow in fluids with low Reynolds numbers, which is important in many industrial applications, such as lubrication, polymer processing, and food processing.
- **Microfluidics:** Stokes's law is used to understand the behavior of particles and fluids in microfluidic systems, which have applications in biotechnology, drug delivery, and lab-on-a-chip devices.
- **Aerosols:** Stokes's law is used to model the behavior of small particles, such as aerosols, in the air. This is important in understanding air pollution, atmospheric chemistry, and the behavior of airborne pathogens.

151. Stoletov's Law

Stoletov's law, also known as the law of photoelectric reciprocity, states that the amount of electric current produced in a photoelectric cell is proportional to the intensity of the incident light, provided that the energy of the photons is greater than the work function of the material in the photoelectric cell. This law is named after the Russian physicist Aleksandr Stoletov, who first formulated it in 1888. The law is one of the fundamental principles underlying the operation of solar cells and other photoelectric devices.

Aleksandr Stoletov

(Credit: https://en.scientificrussia.ru/articles/aleksandr-stoletov-2)

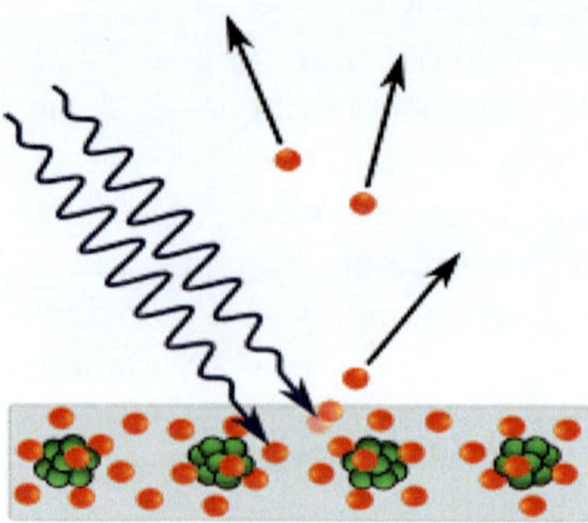

Applications of Stoletov's Law

The law of photoelectric reciprocity provides a fundamental understanding of the interaction between light and materials, allowing for precise control and measurement of light in various applications. The law of photoelectric reciprocity finds applications in several fields. Here are a few examples:

- **Photography:** The law of photoelectric reciprocity is essential in determining proper exposure settings in photography. It helps photographers understand how changes in shutter speed (exposure time) and aperture (light intensity) can affect the overall exposure of an image. By balancing the intensity and duration of light, photographers can achieve desired exposure levels.
- **Photometry:** Photometry involves the measurement of light and its properties. The law of photoelectric reciprocity is employed in photometric measurements to ensure accurate and consistent measurements of light intensity. It helps calibrate photometric instruments and establish standard units of measurement.
- **Radiometry:** Radiometry deals with the measurement of radiant energy, including light. The law of photoelectric reciprocity assists in quantifying the relationship between radiant energy and the resulting photoelectric response. It enables the conversion of light energy to electrical signals in radiometric detectors and sensors.
- **Photobiology:** In the field of photobiology, which studies the effects of light on living organisms, the law of photoelectric reciprocity is relevant for understanding the biological responses to different light intensities and exposure durations. It helps researchers determine optimal conditions for various light-dependent processes, such as photosynthesis, circadian rhythms, and photomorphogenesis.
- **Optoelectronics:** The law of photoelectric reciprocity is also applied in optoelectronic devices that involve the conversion of light energy into electrical signals, such as photodiodes and solar cells. By considering the relationship between light intensity and exposure time, engineers can optimize the design and performance of these devices.

152. SUPERPOSITION PRINCIPLE

The superposition principle is a fundamental concept in physics and engineering that states that when two or more waves or other disturbances interact with each other, the resulting disturbance is the sum of the individual disturbances. In other words, the net effect of the combined disturbances is the algebraic sum of their individual effects.

This principle applies to a wide range of physical phenomena, including sound waves, light waves, electromagnetic fields, and quantum mechanics. For example, when two sound waves of different frequencies and amplitudes overlap, the resulting wave is a combination of the two waves.

The superposition principle is a powerful tool for analyzing complex systems and predicting their behavior. It allows scientists and engineers to break down a complex system into simpler components and analyze each component separately. This is particularly useful in wave mechanics, where the behavior of a wave can be difficult to predict in a complex environment.

Daniel Bernoulli

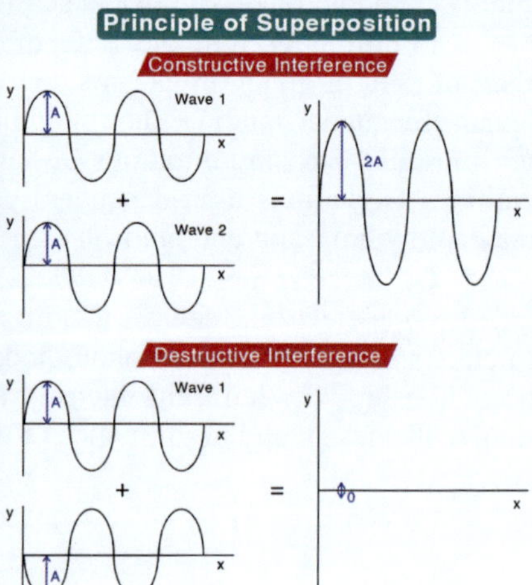

Applications of Superposition Principle

The superposition principle has a wide range of applications in physics, engineering, and other fields, including:

- **Wave interference:** The superposition principle is used to explain wave interference, which occurs when two waves meet and their amplitudes add or cancel out. This phenomenon is important in many fields, including optics, acoustics, and seismology.
- **Circuit analysis:** The superposition principle is used in electrical circuit analysis to determine the behavior of a circuit when multiple sources are present. By considering the effect of each source separately, engineers can analyze and design complex circuits more easily.
- **Quantum mechanics:** The superposition principle is a fundamental concept in quantum mechanics, where it is used to describe the behavior of subatomic particles. According to quantum mechanics, particles can exist in multiple states at the same time, and their wave functions can be combined using the superposition principle.
- **Acoustics:** The superposition principle is used in the design of acoustic spaces, such as concert halls and recording studios. By analyzing the behavior of sound waves and their interactions with surfaces and other objects, engineers can design spaces that optimize sound quality.
- **Image processing:** The superposition principle is used in image processing to combine multiple images into a single composite image. This technique is used in fields such as astronomy and medical imaging to create high-resolution images from multiple sources.
- **Seismology:** The superposition principle is used in seismology to analyze the behavior of seismic waves during earthquakes. By understanding how waves propagate through the earth, scientists can determine the magnitude and location of earthquakes and predict their effects on buildings and infrastructure.

153. Third Law of Thermodynamics

The Third Law of Thermodynamics, also known as the Law of Entropy, states that it is impossible to reach absolute zero temperature. As a system approaches absolute zero temperature, its entropy approaches a minimum value, known as the residual entropy.

The Third Law of Thermodynamics has important implications for the behavior of matter at very low temperatures, such as in the field of cryogenics, and for the study of phase transitions and the properties of materials at the quantum scale. It also helps to explain the behavior of solid materials, such as the formation of defects in crystals, and has important applications in fields such as materials science and condensed matter physics.

Max Planck

Third Law of Thermodynamics

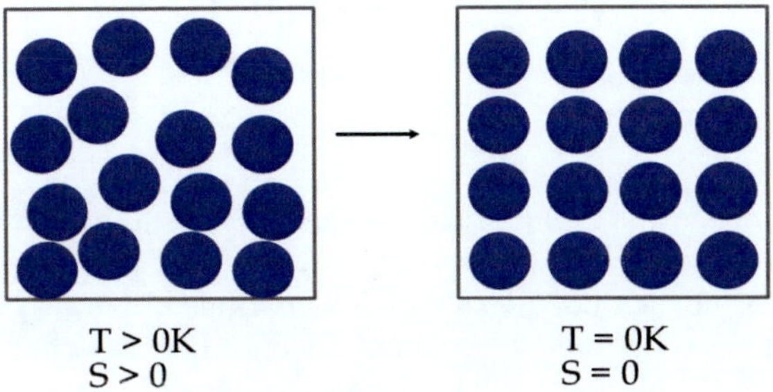

Entropy becomes zero at 0K

Applications of Third Law of Thermodynamics

Third Law of Thermodynamics has a wide range of applications in various fields, and it is an essential part of our understanding of the behavior of matter at very low temperatures and in extreme environments. Some of the applications include:

- **Cryogenics:** The Third Law of Thermodynamics helps us to understand the behavior of materials at very low temperatures. It has important applications in cryogenics, which is the study of the production and behavior of materials at temperatures below -150°C. Cryogenics has many practical applications, including the production of superconductors, which are materials that can conduct electricity without any resistance.
- **Materials science:** The Third Law of Thermodynamics has important applications in the field of materials science, which is concerned with the design, synthesis, and characterization of materials. The law helps us to understand the behavior of materials at very low temperatures, which is important for the development of new materials with specific properties.
- **Chemistry:** The Third Law of Thermodynamics has important applications in chemistry, particularly in the study of phase transitions. Phase transitions occur when a substance changes from one phase to another, such as from a liquid to a solid. The Third Law helps us to understand the behavior of materials at the point of phase transitions.
- **Astrophysics:** The Third Law of Thermodynamics has important implications for the behavior of matter in extreme environments, such as in stars and black holes. The law helps us to understand the behavior of matter under these extreme conditions and provides insights into the behavior of the universe as a whole.

154. TITIUS–BODE LAW

The Titius-Bode law is an empirical law that attempts to predict the spacing between the planets in our solar system. It was first proposed by Johann Titius in 1766 and later popularized by Johann Bode in 1772. The law states that the distances between the planets in our solar system are roughly proportional to the numbers in a mathematical sequence:

0, 3, 6, 12, 24, 48, 96, 192, 384, ...

The first two terms are 0 and 3, which correspond to the distances from the sun to Mercury and Venus, respectively. The next term, 6, is double the distance from the sun to Venus, and corresponds to the distance to Earth. The subsequent terms in the sequence generally match the distances from the sun to Mars, Jupiter, Saturn, Uranus, and Neptune.

The Titius-Bode law is not a true physical law and has no scientific explanation, but it does reflect a pattern in the distribution of the planets. However, the law does not hold true for the distances of Pluto.

Johann Titius

Johann Bode

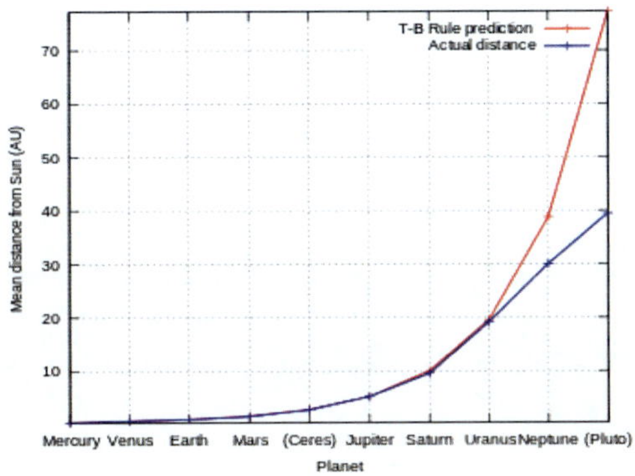

Applications of Titius–Bode Law

The Titius-Bode law is an empirical relationship between the distance of a planet from the sun and its position in the solar system. Although the law is no longer considered scientifically valid, it did lead to the discovery of several new planets.

Johann Daniel Titius and Johann Elert Bode proposed the law in the 18th century. The law states that there is a simple relationship between the distances of the planets from the Sun. Specifically, if we take the sequence of numbers 0, 3, 6, 12, 24, 48, 96, and so on, and add 4 to each number, and then divide the result by 10, we obtain a series of numbers that approximate the average distance of the planets from the Sun in astronomical units (AU). For example, Mercury's distance from the Sun is approximately 0.4 astronomical units (AU), which is close to the value of 0.4 that is obtained by applying the law.

The law is not considered valid because it does not account for the true physical relationships that determine the distances of the planets from the Sun. However, it did lead to the discovery of several new planets. In the late 18th century, astronomers were searching for a planet beyond Saturn. Using the Titius-Bode law as a guide, they predicted that there should be a planet at a distance of 2.8 AU from the Sun. In 1781, Uranus was discovered at a distance of 19.2 AU from the Sun, which was not predicted by the law. However, using the same law, astronomers predicted the existence of a planet at a distance of 38.8 AU from the Sun. In 1846, Neptune was discovered at a distance of 30.1 AU from the Sun. Finally, in the early 20th century, using the Titius-Bode law, astronomers predicted the existence of a planet beyond Neptune. In 1930, Pluto was discovered at a distance of 39.5 AU from the Sun.

155. Torricelli's Law

Torricelli's law is a principle in fluid dynamics that describes the velocity of a fluid flowing out of an opening in a container. It is named after Italian physicist and mathematician Evangelista Torricelli, who first described the principle in the 17th century.

According to Torricelli's law, the velocity of a fluid flowing out of an opening in a container is proportional to the square root of the height of the fluid above the opening. This can be expressed mathematically as:

$$v = (2gh)^{1/2}$$

where v is the velocity of the fluid, g is the acceleration due to gravity, and h is the height of the fluid above the opening.

Torricelli's law is often used to calculate the velocity of water flowing out of a spout or the pressure of fluid in a container. It is also a key principle in the design of many fluid systems, including pumps, pipes, and hydraulic machinery.

Evangelista Torricelli

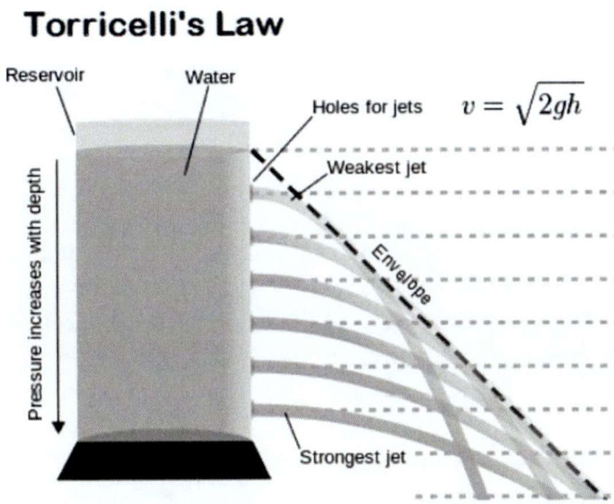

Applications of Torricelli's Law

Torricelli's law, also known as Torricelli's theorem, relates the speed of a fluid flowing through a small hole in a container to the height of the fluid above the hole. The law has several applications, including:

- **Measuring fluid flow rate:** Torricelli's law can be used to measure the flow rate of fluids in pipes or channels. By measuring the height of the fluid above a small hole in the pipe, the flow rate can be calculated.
- **Designing fountains:** Torricelli's law is used to design fountains, especially those that eject water in a parabolic trajectory. By controlling the height of the fluid above the nozzle, the velocity and trajectory of the water can be controlled.
- **Designing hydraulic jumps:** Hydraulic jumps occur when a high-velocity liquid abruptly slows down, resulting in a rapid increase in pressure and turbulence. Torricelli's law is used to design hydraulic jumps in channels and spillways, ensuring that the jump occurs at the desired location.
- **Vacuum generation:** Torricelli's law is the basis of vacuum generation in laboratory and industrial settings. By connecting a container of fluid to a vacuum pump, the pump can remove air from the container, creating a partial vacuum.

156. Umov Effect

The Umov effect, also known as Umov's law, is a relationship between the reflectivity (or geometric albedo) of an object and the degree of linear polarization of light scattered by it. Albedo is the fraction of sunlight that is diffusely reflected by a body. It is measured on a scale from 0 (corresponding to a black body that absorbs all incident radiation) to 1 (corresponding to a body that reflects all incident radiation). The effect was discovered by the Russian physicist Nikolay Umov in 1905 and can be observed for celestial objects such as the surface of the Moon and the asteroids.

The degree of linear polarization of light P is defined by

$$\mathbf{P = (I_N - I_P) / (I_N + I_P)}$$

where I_N and I_P are the intensities of light in the directions Normal (perpendicular) and parallel to the plane of a polarizer aligned in the plane of reflection. Values of P are zero for unpolarized light, and ±1 for linearly polarized light.

Umov's law states

$$\mathbf{P \propto 1 / \alpha}$$

where α (or A) is the albedo of the object. Thus, highly reflective objects tend to reflect mostly unpolarized light, and dimly reflective objects tend to reflect polarized light. The law is only valid for large phase angles (angles between the incident light and the reflected light).

Nikolay Umov

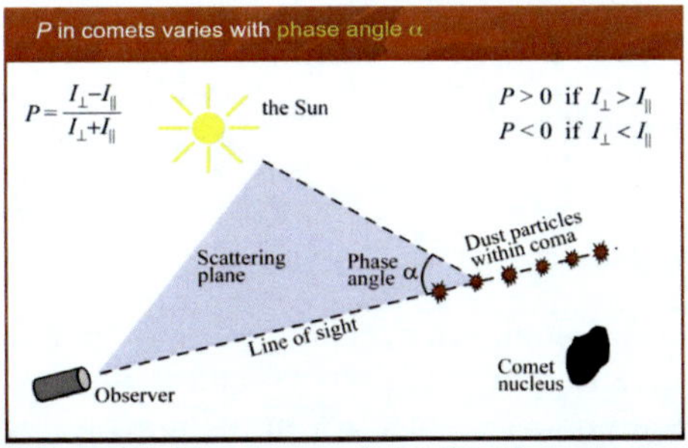

The Umov Effect (Zubko, Evgenij. 2012. Remote Sensing of Comets Using Polarimetry. Accessed February 22, 2012)

Applications of Umov Effect

The relationship between reflectivity and the degree of linear polarization of scattered light has diverse applications in fields ranging from planetary science and remote sensing to material analysis and optical engineering. Umov's law has several applications in various fields.

- **Remote sensing and planetary science:** In remote sensing, the measurement of the degree of linear polarization of scattered light can provide valuable information about the surface properties of objects such as planets, moons, asteroids, and comets. By studying the relationship between reflectivity and polarization, scientists can gain insights into surface composition, texture, and structure of these celestial bodies. This information helps in characterizing planetary surfaces, understanding geological processes, and identifying different surface materials.
- **Material analysis and characterization:** The polarization characteristics of scattered light can be utilized for material analysis and characterization. By studying the polarization properties of light reflected or scattered from surfaces, scientists can identify and distinguish different types of materials, such as minerals, biological tissues, or manufactured objects. This has applications in fields like material science, geology, and biomedicine.
- **Remote sensing of earth's atmosphere:** The relationship between reflectivity and the degree of linear polarization is also useful in atmospheric remote sensing. By analyzing the polarization characteristics of scattered light from the atmosphere, researchers can retrieve information about aerosols, cloud properties, and atmospheric composition. This aids in climate studies, weather forecasting, and environmental monitoring.

157. Van der Waals Equation

The Van der Waals equation is an equation of state for a real gas that takes into account the effects of intermolecular forces and the finite size of gas molecules, which are not accounted for in the ideal gas law. The Van der Waals equation is named after the Dutch physicist Johannes Diderik van der Waals, who derived it in 1873.

The Van der Waals equation is given by:

$$\left[P + \frac{an^2}{v^2}\right](V - nb) = nRT$$

where P is the pressure, V is the volume, n is the number of moles of the gas, R is the ideal gas constant, T is the temperature, a is a measure of the intermolecular forces between the gas molecules, and b is a measure of the volume excluded by the gas molecules.

The Van der Waals equation is a more accurate model of real gases than the ideal gas law, particularly at high pressures and low temperatures. The term (n/V)^2 accounts for the attractive forces between the gas molecules, while the term nb accounts for the finite size of the gas molecules. The parameter a is a measure of the strength of the attractive forces between the gas molecules, while b is a measure of the volume excluded by the gas molecules.

Johannes Diderik van der Waals

$$PV = nRT \longrightarrow \left(P + \frac{an^2}{V^2}\right)(V - nb) = nRT$$

Correction for molecular attraction

Correction for volume of molecules

Applications of Van der Waals Equation

The Van der Waals equation is a more accurate model of real gases than the ideal gas law, and it has a number of applications in various fields of science and engineering, particularly in the study of phase transitions and critical phenomena. It is used to model the behavior of real gases and to predict the conditions under which a gas will undergo a phase transition, such as condensation or liquefaction. The Van der Waals equation is also used in the study of critical phenomena, which are the phenomena that occur at the critical point of a phase transition. Here are some applications of the Van der Waals equation.

- **Design of refrigeration systems:** The Van der Waals equation is used in the design of refrigeration systems, where gases are used as refrigerants. It is used to calculate the temperature and pressure conditions required for the refrigerant to undergo phase transitions, such as condensation or evaporation.
- **Study of critical phenomena:** The Van der Waals equation is used in the study of critical phenomena, which occur at the critical point of a phase transition. It is used to predict the behavior of gases near the critical point, and to determine the critical exponents that describe the behavior of the system near the critical point.
- **Prediction of phase behavior:** The Van der Waals equation is used to predict the behavior of gases at high pressures and low temperatures, where the ideal gas law breaks down. It is used to predict the conditions under which gases will undergo phase transitions, such as liquefaction or condensation.
- **Design of gas storage and transport systems:** The Van der Waals equation is used to design gas storage and transport systems. It is used to calculate the conditions under which gases will remain in the gaseous phase, and to determine the volume and pressure required to store or transport a given amount of gas.
- **Petroleum engineering:** The Van der Waals equation is used to predict the conditions under which natural gas will condense or liquefy in the reservoir, and to determine the amount of gas that can be extracted.

158. VLASOV EQUATION

The Vlasov equation is a partial differential equation that describes the evolution of a distribution function of a system of charged particles under the influence of their self-consistent electromagnetic fields. It is named after Anatoly Vlasov, who introduced it in 1938. The Vlasov equation is a fundamental equation in plasma physics and is used to study the dynamics of plasmas, which are ionized gases consisting of charged particles such as electrons and ions. The Vlasov equation can be derived from the Boltzmann equation in the limit of weak collisionality. It describes the time evolution of the distribution function of particles in phase space, which specifies the probability density of finding a particle with a particular position and momentum at a given time. The Vlasov equation is a nonlinear equation and is generally solved numerically using computational techniques.

Vlasov with the collisionless Boltzmann equation (called the Vlasov equation), in generalized coordinates is:

$$\mathbf{d/dt \cdot f\,(r, p, t\,) = 0}$$

and explicitly a partial differential equation is:

$$\frac{\partial f}{\partial t} + \frac{d\vec{r}}{dt} \cdot \frac{\partial f}{\partial \vec{r}} + \frac{d\vec{p}}{dt} \cdot \frac{\partial f}{\partial \vec{p}} = 0$$

Anatoly Vlasov

(Credit: https://alchetron.com/Anatoly-Vlasov)

Applications of Vlasov Equation

The Vlasov equation is a partial differential equation that describes the evolution of the distribution function of a plasma or a gas. It is widely used in plasma physics to model a variety of phenomena. Some applications of the Vlasov equation include:

- **Kinetic theory of plasma:** The Vlasov equation provides a fundamental framework for understanding the behavior of plasma particles. It is used to study the evolution of the distribution function of electrons and ions in a plasma under the influence of external fields and collisions.
- **Plasma waves:** The Vlasov equation is used to study the propagation of plasma waves, which are collective oscillations of plasma particles. Examples of plasma waves include Langmuir waves, ion acoustic waves, and electromagnetic waves.
- **Magnetic confinement fusion:** The Vlasov equation is used to model the behavior of plasma in magnetic confinement devices, such as tokamaks and stellarators. It is used to study the stability of plasma, the formation of plasma currents, and the transport of particles and energy.
- **Space plasmas:** The Vlasov equation is used to study the behavior of plasma in the Earth's magnetosphere and the solar wind. It is used to understand the generation of plasma waves and the transport of energetic particles in these regions.
- **Astrophysical plasmas:** The Vlasov equation is used to study the behavior of plasma in astrophysical systems, such as the interstellar medium, accretion disks, and supernova remnants. It is used to understand the formation of shocks, the acceleration of particles, and the generation of magnetic fields.

159. WAVE-PARTICLE DUALITY

Wave-particle duality is a fundamental concept in quantum mechanics that describes the behavior of matter and light. It is the idea that particles, such as electrons, can exhibit wave-like properties and waves, such as light, can exhibit particle-like properties.

The wave-like behavior of particles is most evident in phenomena such as diffraction and interference, where particles can exhibit patterns similar to those produced by waves. For example, when electrons are passed through a double-slit experiment, they can exhibit an interference pattern that is characteristic of waves.

The particle-like behavior of waves is most evident in phenomena such as the photoelectric effect, where light can be thought of as particles (photons) that transfer energy to electrons, causing them to be ejected from a material.

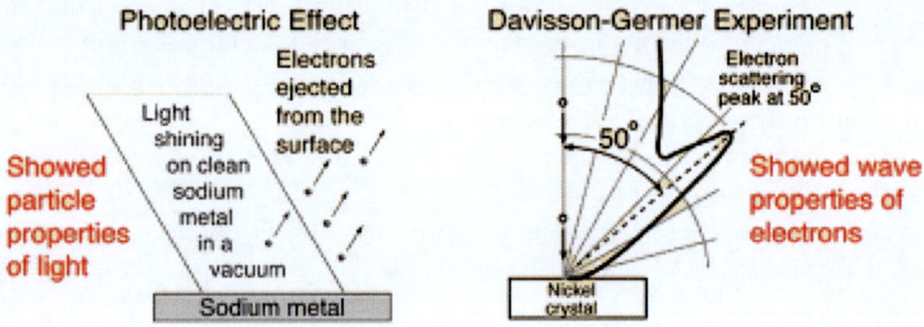

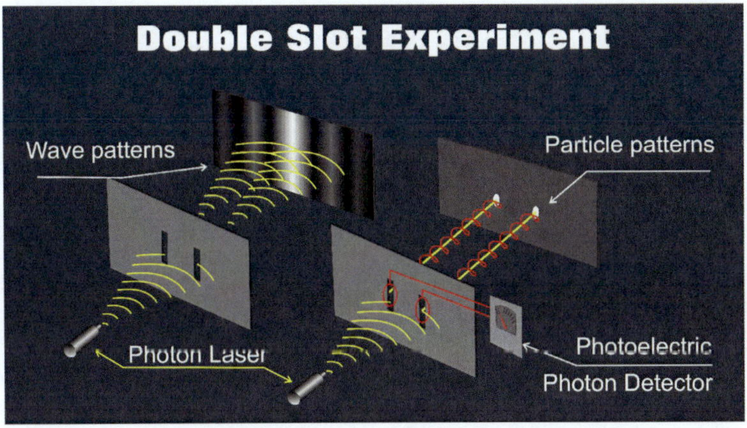

Applications of Wave-Particle Duality

Concept of wave-particle duality is a fundamental concept in modern physics that describes the behavior of matter and light. It has numerous applications in various fields of physics and engineering, and it is essential for understanding the behavior of particles and waves at the quantum level. Its applications include:

- **Quantum mechanics:** The wave-particle duality is a fundamental concept in quantum mechanics and underlies many of its predictions and theories.
- **Particle physics:** The wave-particle duality is used to describe the behavior of subatomic particles in particle physics experiments.
- **Solid state physics:** The wave-particle duality is used to understand the behavior of electrons in solids, which is important in the design of electronic devices.
- **Optics:** The wave-particle duality is used to describe the behavior of light in optical systems, which is important in the design of optical instruments and communication systems.

160. Weinberg–Witten Theorem

The Weinberg-Witten theorem is a result in theoretical physics that establishes a connection between the electroweak force and gravity. Specifically, the theorem states that any consistent theory of quantum gravity must incorporate the full Standard Model of particle physics, including the electroweak force.

The theorem was first proposed by Steven Weinberg and Edward Witten in the 1980s. It has important implications for the search for a unified theory of physics, which would bring together all of the fundamental forces in the universe into a single framework. It suggests that any such theory must include both the electroweak force and gravity, and that these two forces must be intimately connected in some way.

The proof of the Weinberg-Witten theorem is quite technical and relies on a number of advanced mathematical concepts. It is generally considered to be one of the most important results in modern theoretical physics, as it provides a powerful constraint on the development of any future theories of physics.

> In theoretical physics, the Weinberg–Witten theorem states that massless particles (either composite or elementary) with spin $j > 1/2$ cannot carry a Lorentz-covariant current, while massless particles with spin $j>1$ cannot carry a Lorentz-covariant stress-energy. The theorem is usually interpreted to mean that the graviton ($j=2$) cannot be a composite particle in a relativistic quantum field theory.

Steven Weinberg **Edward Witten**
(Credit: https://citaty.net/autori/steven-weinberg/)

Applications of Weinberg–Witten Theorem

The Weinberg-Witten theorem is a result in theoretical particle physics that relates the masslessness of the photon and the existence of gauge invariance in a quantum field theory. It states that any consistent quantum field theory of massless spin-1 particles must have a gauge symmetry. This theorem has important implications in the development of theories of elementary particles and their interactions.

- One important application of the Weinberg-Witten theorem is in the development of the standard model of particle physics. The standard model describes the interactions of elementary particles, including the electromagnetic force mediated by photons. The masslessness of the photon is a fundamental property of this force, and the existence of gauge invariance is a crucial component of the theory.
- Another application of the Weinberg-Witten theorem is in the development of theories of quantum gravity. The theorem suggests that a consistent theory of gravity must also have a gauge symmetry, which has led to the development of theories such as string theory and loop quantum gravity.

Weinberg-Witten theorem has played a significant role in the development of modern theoretical physics, providing a fundamental understanding of the relationship between gauge symmetry and the masslessness of certain elementary particles.

161. WIEDEMANN-FRANZ LAW

The Wiedemann-Franz Law is a fundamental principle in physics that describes the relationship between the electrical conductivity and thermal conductivity of a material. It was named after the physicists Gustav Wiedemann and Rudolf Franz, who first proposed the relationship in 1853.

The Wiedemann-Franz Law states that the ratio of the electrical conductivity to the thermal conductivity of a material is proportional to the absolute temperature, and is given by:

$$L = k / \sigma T$$

where L is the Lorenz number, k is the thermal conductivity, σ is the electrical conductivity, and T is the absolute temperature. According to the Wiedemann-Franz Law, as the temperature of a material increases, the ratio of the electrical conductivity to the thermal conductivity also increases.

Gustav Wiedemann

$$\frac{\kappa}{\sigma} = LT \quad or \quad L = \frac{\kappa}{\sigma T} \qquad \text{Wiedemann-Franz Law}$$

κ = thermal conductivity $\quad$ σ = electrical conductivity

L = Lorenz number

Applications of Wiedemann-Franz Law

The Wiedemann-Franz law is important for understanding the behavior of materials in a variety of applications, such as thermoelectric devices, which convert heat into electricity. It is also used in the study of metals and semiconductors, and in the development of new materials for electronic and energy applications. The Wiedemann-Franz Law has several applications in materials science, engineering, and technology. Some of the most important applications are:

- **Thermoelectric materials:** The Wiedemann-Franz Law is used to design and optimize thermoelectric materials, which can convert heat into electricity. By controlling the ratio of electrical conductivity to thermal conductivity, researchers can develop materials with high thermoelectric efficiency.
- **Heat management:** The Wiedemann-Franz Law is used in the design of heat sinks and other cooling systems, which are used to dissipate heat from electronic devices. By understanding the relationship between the electrical and thermal properties of materials, engineers can develop more efficient cooling systems.
- **Semiconductor technology:** The Wiedemann-Franz Law is used to optimize the performance of semiconductor devices, such as transistors and diodes. By controlling the electrical and thermal properties of materials, researchers can develop semiconductors with improved efficiency and reliability.
- **Materials characterization:** The Wiedemann-Franz Law is used to measure the electrical and thermal properties of materials, which are important for understanding their behavior in different conditions. By measuring the ratio of electrical conductivity to thermal conductivity, researchers can determine the properties of materials, such as their band gap, doping level, and defect density.

Wiedemann-Franz Law is a fundamental principle in materials science and engineering, which provides a useful framework for understanding the relationship between the electrical and thermal properties of materials. Its applications are diverse and far-reaching, from developing more efficient thermoelectric materials to optimizing the performance of semiconductor devices.

162. Wien's Law

Wien's law also called Wien's displacement law is a physical law that relates the wavelength of the peak emission of a blackbody to its temperature. The law is named after the German physicist Wilhelm Wien, who discovered it in 1893.

The statement of Wien's law is: "The wavelength of the peak of the blackbody radiation curve is inversely proportional to the temperature of the body."

In mathematical terms, Wien's law is expressed as:

$$\lambda_{max} = b / T$$

where λ_{max} is the wavelength of the peak emission, T is the temperature of the blackbody, and b is Wien's constant, which has a value of approximately 2.898 x 10^-3 m.K.

Wien's law is important in astrophysics and astronomy, where it is used to determine the temperature of stars and other celestial objects based on the wavelength of their peak emission. It is also used in materials science and engineering, where it is used to study the thermal radiation properties of materials and to design and optimize thermophotovoltaic devices for energy conversion.

Wilhelm Wien

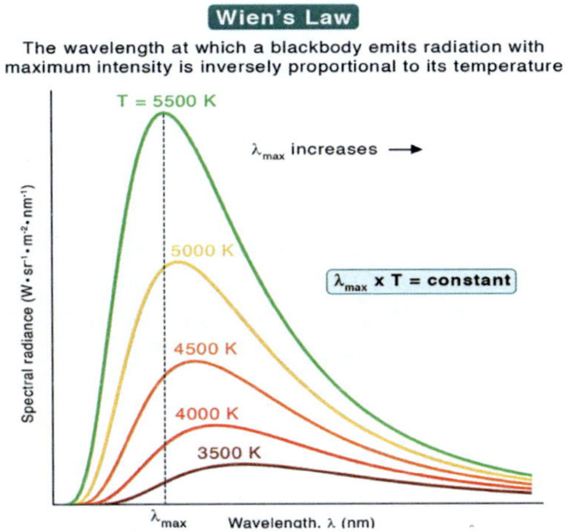

Applications of Wien's Law

Wien's law has various applications in different fields of science and engineering. Some of the significant applications are:

- **Astrophysics and astronomy:** Wien's law is extensively used in astrophysics and astronomy to determine the temperature of celestial objects, such as stars, galaxies, and planets. By analyzing the peak wavelength of their radiation, astronomers can estimate the temperature and other properties of these objects.
- **Materials science and engineering:** Wien's law is also used in the field of materials science and engineering to design and optimize thermophotovoltaic devices that convert thermal radiation to electricity. By choosing materials that have a high emissivity and a narrow spectral bandwidth of emission, the efficiency of such devices can be enhanced.
- **Infrared spectroscopy:** Infrared spectroscopy is a powerful analytical technique that uses the absorption of infrared radiation to identify and characterize chemical compounds. Wien's law is used to determine the temperature of the sample and to calculate the wavelength range of the infrared radiation that will be absorbed by the sample.
- **Thermal imaging:** Thermal imaging is a technique that uses infrared radiation to create images of objects based on their temperature. Wien's law is used to determine the temperature range and the optimal wavelength range for thermal imaging sensors.

163. YOUNG–LAPLACE EQUATION

The Young-Laplace equation is a partial differential equation that relates the curvature of a liquid interface with the pressure difference across the interface. It is named after Thomas Young and Pierre-Simon Laplace, who independently derived it in the early 19th century.

The Young-Laplace equation is given by:

$$\Delta P = \gamma\,(1/R_1 + 1/R_2)$$

where ΔP is the pressure difference across the interface, γ is the surface tension of the liquid, and R_1 and R_2 are the principal radii of curvature of the interface. If the interface is a flat plane, the curvature is zero and the pressure difference is also zero. However, if the interface is curved, the pressure difference is proportional to the surface tension and the curvature of the interface.

The Young-Laplace equation is used to study a variety of phenomena related to liquid interfaces, such as capillary action, bubble formation and stability, and wetting of surfaces by liquids. It is also used in engineering applications, such as the design of microfluidic devices and the study of fluid flow in porous media.

Thomas Young **Pierre-Simon Laplace**

(Credit: Portrait by Henry Briggs; Copyright © The Royal Society, http://rstb.royalsocietypublishing.org/content/370/1666/20140308)

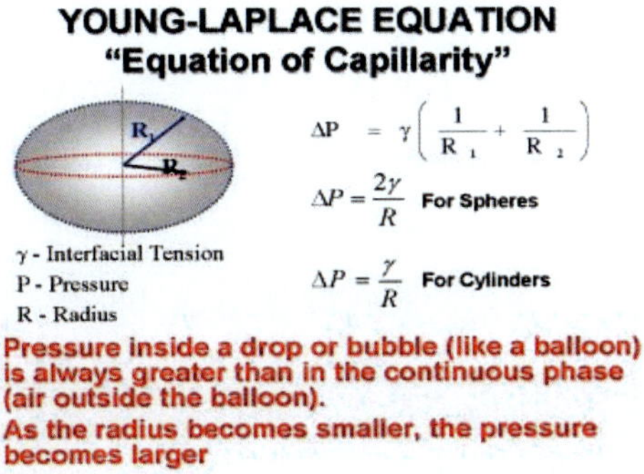

Applications of Young–Laplace Equation

The Young-Laplace equation is a fundamental equation in fluid mechanics that describes the curvature of an interface between two fluids in contact. Some of its applications are:

♦ **Capillary action:** The Young-Laplace equation is used to describe the rise or fall of a liquid in a narrow tube due to capillary action. The equation relates the curvature of the liquid surface to the surface tension and the pressure difference across the curved surface.

♦ **Bubble formation and stability:** The Young-Laplace equation is used to study the formation and stability of bubbles in a liquid. The equation relates the pressure inside the bubble to the radius of the bubble and the surface tension of the liquid.

♦ **Wetting and adhesion:** The Young-Laplace equation is used to study the wetting and adhesion of liquids on solid surfaces. The equation relates the contact angle between the liquid and the solid surface to the surface tension of the liquid and the curvature of the liquid surface near the contact line.

♦ **Stability of emulsions and foams:** The Young-Laplace equation is used to study the stability of emulsions and foams, which are mixtures of immiscible liquids and gases, respectively. The equation relates the curvature of the interface between the two phases to the surface tension and the pressure difference across the interface.

♦ **Fluid dynamics:** The Young-Laplace equation is used in the study of fluid dynamics to describe the shape and motion of fluid interfaces in a variety of settings, such as in the flow of fluids through porous media, in the breakup of liquid droplets, and in the deformation of bubbles and drops by fluid flows.

164. ZEROTH LAW OF THERMODYNAMICS

The Zeroth Law of Thermodynamics is a fundamental principle of thermodynamics that states that if two bodies are each in thermal equilibrium with a third body, then they are also in thermal equilibrium with each other. This principle establishes the concept of temperature and allows the construction of a temperature scale.

The Zeroth Law is called "zeroth" because it was discovered after the first and second laws of thermodynamics had already been established. The law was added to the laws of thermodynamics to clarify the fundamental concepts of thermal equilibrium and temperature.

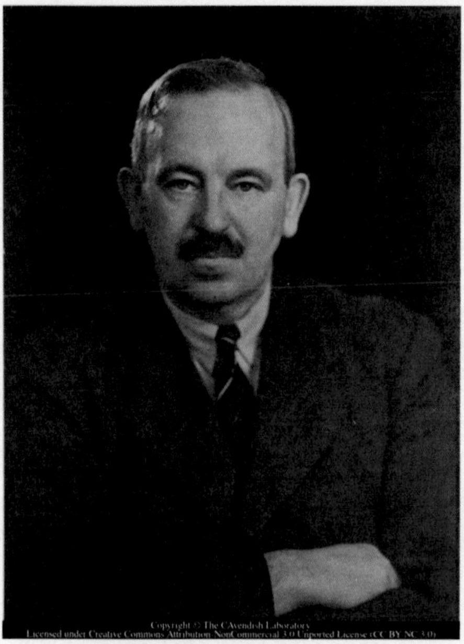

Ralph Howard Fowler

(Credit: https://images.lib.cam.ac.uk/content/images/PH-CAVENDISH-P-00401-000-00001.jpg)

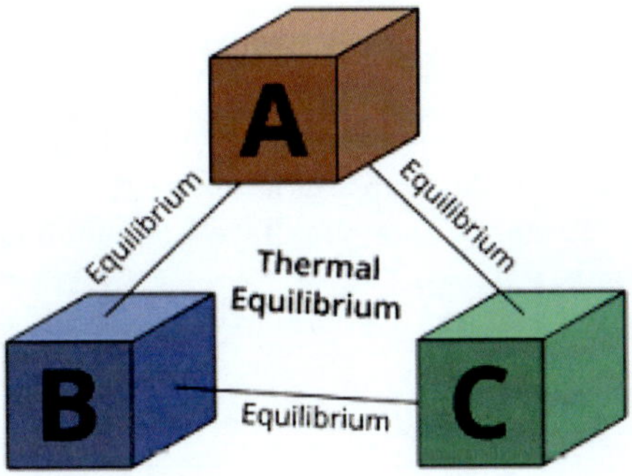

Zeroth law of Thermodynamics

Applications of Zeroth Law of Thermodynamics

Zeroth Law of Thermodynamics is a fundamental principle of thermodynamics that establishes the concept of thermal equilibrium and temperature. It has numerous practical applications in areas such as thermometer calibration, temperature control, heat transfer, and food processing. The Zeroth Law of Thermodynamics has numerous practical applications, including:

- **Calibration of thermometers:** The law is used to calibrate thermometers, which are used to measure temperature in a wide range of applications. Thermometers are calibrated by comparing them to a known reference temperature using the Zeroth Law.
- **Temperature control:** The law is used in temperature control systems, such as thermostats, which regulate temperature by comparing the temperature of a system to a setpoint temperature.
- **Heat transfer:** The law is used in the study of heat transfer, which is the transfer of thermal energy from one system to another. The law establishes the concept of thermal equilibrium, which is necessary for the transfer of heat between two systems.
- **Food processing:** The law is used in the food industry to ensure food safety and quality. The law is applied to determine the minimum and maximum temperatures required to kill bacteria and other harmful microorganisms in food processing.

REFERENCES

Burg, Jean-Pierre. 2018. “Brittle Faulting”. ETH Library, Script to Structural Geology, 53-96. https://doi.org/10.3929/ethz-b-000279493.

Dronkers, J. 2022. “Bruun rule for shoreface adaptation to sea-level rise.” Accessed December 11, 2022. http://www.coastalwiki.org/wiki/Bruun_rule_for_shoreface_adaptation_to_sea-level_rise.

Leavitt, Henrietta. S, Pickering, Edward C. 1912. "Periods of 25 Variable Stars in the Small Magellanic Cloud". Harvard College Observatory Circular 173: 1–3. Bibcode:1912HarCi.173....1L.

Nolte, D. David. 2019. “How to Teach General Relativity to Undergraduate Physics Majors”. Accessed July 15, 2019. https://galileo-unbound.blog/2019/07/15/how-to-teach-general-relativity-to-undergraduate-physics-majors.

Sonium. 2015. “Boltzmann distribution graph”. Accessed March 30, 2015. https://commons.wikimedia.org/wiki/File:Boltzmann_distribution_graph.svg.

Strobel, Nick. 2001. “Curved Spacetime”. Accessed May 28, 2001. https://www.astronomynotes.com/relativity/s3.htm.

Strobel, Nick. 2022. “Neutrino”. Accessed June 19, 2022. https://www.astronomynotes.com/starsun/s4.htm.

Vanstone, Emma. 2022. “What is Hooke’s Law?”. Accessed April 20, 2022. https://www.science-sparks.com/what-is-hookes-law.

Yvon, Jacques. 1935. “La théorie statistique des fluides et l’équation d’état (in French)”. Volume 203 of Actualités scientifiques et industrielles. ISSN 0365-6861. Paris: Hermann & cie.

Zubko, Evgenij. 2012. Remote Sensing of Comets Using Polarimetry. Accessed February 22, 2012. https://www.cps-jp.org/~cps/pub/seminar/fy2011/2012-02-22/zubko/pub/20120222_zubko_01.pdf.